Code of Practice
for Electric Vehicle Charging Equipment Installation

3rd Edition

Published by The Institution of Engineering and Technology, London, United Kingdom

The Institution of Engineering and Technology is registered as a Charity in England & Wales (no. 211014) and Scotland (no. SC038698).

First published 2012 (978-1-84919-514-0)
Second edition 2015 (978-1-84919-839-4)
Third edition 2018 (978-1-78561-680-8)

The Institution of Engineering and Technology
Michael Faraday House
Six Hills Way, Stevenage
Herts, SG1 2AY, United Kingdom

www.theiet.org

While the publisher, author and contributors believe that the information and guidance given in this work is correct, all parties must rely upon their own skill and judgement when making use of it. Neither the publisher, nor the author, nor any contributors assume any liability to anyone for any loss or damage caused by any error or omission in the work, whether such error or omission is the result of negligence or any other cause. Any and all such liability is disclaimed.

A list of organisations represented on this committee can be obtained on request to IET standards. This publication does not purport to include all the necessary provisions of a contract. Users are responsible for its correct application. Compliance with the contents of this document cannot confer immunity from legal obligations.

It is the constant aim of the IET to improve the quality of our products and services. We should be grateful if anyone finding an inaccuracy or ambiguity while using this document would inform the IET standards development team, ietstandardsenquiries@theiet.org, The IET, Six Hills Way, Stevenage SG1 2AY, UK.

ISBN 978-1-78561-680-8 (paperback)
ISBN 978-1-78561-681-5 (electronic)

Contents

Acknowledgements

The IET would like to acknowledge the following organisations for their contributions to this document.

Lead author – Graham Kenyon (G Kenyon Technology Ltd)

Technical Committee
Atkins
BMM Energy Solutions
British Electrotechnical and Allied Manufacturers Association (BEAMA)
British Gas
Cenex
Chargemaster plc
EA Technology
Electrical Contractors Association (ECA)
Electrical Safety First
Energy Networks Association (ENA)
Energy Technologies Institute (ETI)
E.ON
Ford Motor Company
Health & Safety Executive (HSE)
Institution of Engineering and Technology
John Dallimore and Partners
Lyra Electronics
Ministry of Housing, Communities and Local Government (MHCLG)
Office for Low Emission Vehicles (OLEV)
Qarma Solutions Ltd
Society of Motor Manufacturers and Traders (SMMT)
Transport Research Laboratory Ltd (TRL)
Zero Carbon Futures

Foreword

Electric vehicles have an important role to play in meeting air quality legislation and the UK's commitment to climate change targets. For these reasons, the UK government is actively supporting the switch to electric vehicles.

Most of the charging needs are satisfied by dedicated BS 1363 socket-outlets with charging limited to 10 A. This low-power AC charging solution meets the minimum charging needs of the average journey length of 25 miles daily within a charging time of approximately three hours.

For many users, with the increase in battery size of electric vehicles, considerably more energy can be stored than the average daily commute requires. A large electric vehicle parking facility creates a major opportunity to provide battery storage capacity to the electricity system, mitigating the intermittence of renewable generation and network constraints. Other users need a larger range over which a vehicle can travel on a single charge, with a shorter time taken to recharge the vehicle. A greater range requires a larger battery pack, increasing the time taken to recharge the battery using a standard BS 1363 type socket-outlet. To address this issue, vehicle manufacturers have worked with the electrical industry to develop suitable solutions.

The solutions that have been developed involve dedicated electric vehicle charging equipment using dedicated socket-outlets, plugs and vehicle connectors. They also provide additional safety by requiring communication between the vehicle and the charging equipment before charging can commence. Two conductive charging solutions are now widely supported by the automotive industry:

(a) AC charging using the vehicle on-board charger; and
(b) DC charging using external chargers provided by the electric vehicle supply equipment (EVSE).

Note: This Code of Practice has been amended to include the requirements of BS 7671:2018 *Requirements for Electrical Installations*.

SECTION 1

Scope

This Code of Practice applies to the installation of dedicated conductive charging equipment for the charging of pure electric and plug-in hybrid electric road vehicles (PHEV), including extended range electric vehicles (E-REV).

The installation of wireless power transfer (WPT) charging equipment is not currently covered in this Code of Practice. This will be reviewed in future editions, once WPT technology has matured and more experience has been gained with its installation. Some guidance based on the information currently available is provided in Annex F.

The Code of Practice covers the installation of dedicated electric vehicle charging equipment of the types described in Section 4 of the Code. It is not intended to cover socket-outlets provided for general power; references to socket-outlets in this Code are either to dedicated electric vehicle charging socket-outlets or to standard socket-outlets incorporated into dedicated electric vehicle charging equipment.

It covers the installation of both AC and DC charging equipment intended for plug-in electric vehicles (PEV) complying with the BS EN 61851 series of standards.

This edition of the Code of Practice:

(a) discusses vehicles used for electrical energy storage (in conjunction with the IET *Code of Practice for Electrical Energy Storage Systems*) – see Section 10; and
(b) introduces integration with smart metering and control, automation and monitoring systems – see Section 12.

It applies to the installation of conductive electric vehicle charging equipment in all locations where such equipment may be required, including:

(a) domestic installations, such as installations in, or adjacent to, houses and their associated garages;
(b) on-street installations; and
(c) commercial and industrial installations, such as installations in, or adjacent to, business premises, e.g. shops, offices, factories, etc., including public and private car parks, whether single-level or multi-storey, and filling stations.

Note: Standards for WPT installations are still under development. A commentary is included in Annex F.

The objective of this Code of Practice is to provide guidance on the installation of electric vehicle charging equipment to assist the installer in ensuring that the final installation complies with the relevant requirements of BS 7671:2018 (The IET Wiring Regulations, 18th Edition) and, where necessary, the Electricity Safety, Quality and Continuity Regulations (ESQCR) 2002 (as amended).

Whilst this Code of Practice addresses the technical aspects of electric vehicle charging equipment installation, it does not specifically address the local authority planning regulations and/or building regulations that may need to be considered when installing such equipment.

SECTION 2

Overview of electric vehicle charging equipment

2.1 Introduction

Electric vehicle supply equipment (EVSE) is a key component in the roll-out of charging infrastructure for electric vehicles using a conductive or wireless charging solution.

Most electric vehicles can charge from a conventional three-pin socket using the mode 2 charging equipment often supplied with the vehicle. Increasingly, more sophisticated and digitally-managed EVSE is being deployed in the residential, business and public infrastructure context, providing charging management capabilities such as access control, configurable charging power and accountability for charging events.

EVSE for conductive charging solutions supports two principal charging options:

(a) AC supply for the on-board vehicle charger; and
(b) EVSE charger for DC supply to the vehicle battery.

There is a wide power range in EVSE, which may range from as low as 6 A to well over 100 A. This can create significant challenges, not just from an EVSE installation perspective, but also in terms of integration and management as part of the electricity supply system.

EVSE increasingly forms part of onsite generation and energy storage solutions, enabling the vehicle to act as a mobile energy storage device that can be charged as well as discharged via the charging infrastructure. This adds additional requirements to be considered as part of the EVSE installation design and its operational characteristics.

The BS EN 61851 series of standards specifies the design and performance requirements for electric vehicle conductive charging equipment with which all equipment shall comply. Furthermore, electric vehicle charging equipment shall also comply with the Electromagnetic Compatibility Regulations 2016 and the Electrical Equipment (Safety) Regulations 2016, and be CE-marked accordingly.

BS 7671:2018 is the appropriate safety standard for electrical installation.

2.2 Charging equipment for Modes 1, 2, 3 and 4

There are four commonly adopted charging methods, called 'modes'. These are described in 2.2.1.

2.2.1 Mode 1 charging

In Mode 1 charging, connection of the electric vehicle to the AC supply network utilizes standardized socket-outlets not exceeding 16 A and not exceeding 250 V AC single-phase or 480 V AC three-phase, at the supply side, and utilizing the power and protective earth conductors (according to BS EN 61851-1). Refer to Figure 2.1 below.

Note: Mode 1 is no longer a mainstream charging technology.

▼ **Figure 2.1 Mode 1 charging**

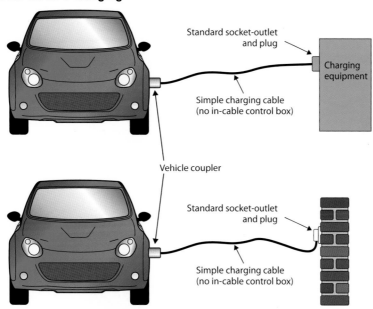

2.2.2 Mode 2 charging

Mode 2 charging describes the minimal charging solution for single-phase domestic socket-outlets. It provides charging currents of 10 A or less. In Mode 2 charging, connection of the EV to the AC supply network utilizes standardized socket-outlets not exceeding 32 A and not exceeding 250 V AC single-phase or 480 V AC three-phase, at the supply side, and utilizes the power and protective earth conductors together with a control pilot function and system of personnel protection against electric shock (BS EN 62752) between the EV and the in-cable control box (ICCB). Refer to Figure 2.2 below.

▼ **Figure 2.2 Mode 2 charging**

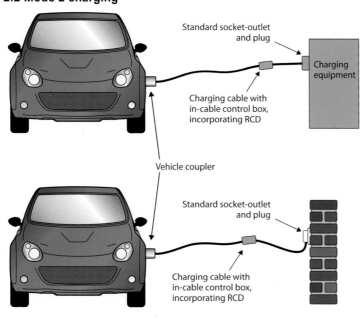

2.2.3 Mode 3 charging

In Mode 3 charging, connection of the EV to the AC supply network utilizes dedicated EVSE where the control pilot function extends to control equipment in the EVSE, permanently connected to the AC supply network. The EVSE may be supplied from a three-phase AC supply and will often incorporate BS EN 62196-2 Type 2 plugs where tethered cables are not used. Refer to Figures 2.3 and Figure 2.4 below.

▼ **Figure 2.3 Mode 3 charging (dedicated socket-outlet)**

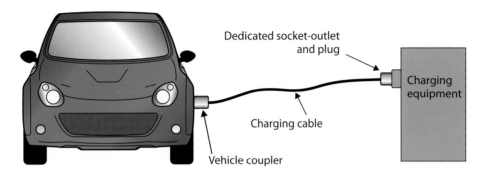

▼ **Figure 2.4 Mode 3 charging (tethered charging cable)**

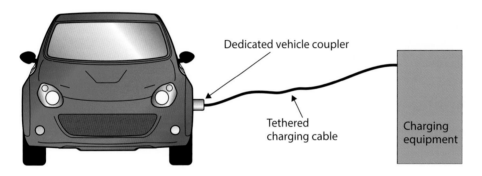

AC charging equipment tends to be either in the form of 'charging posts' or wall-mounted charging units, and comes in a variety of current ratings, for example, 13 A, 16 A and 32 A. Refer to Figure 2.5 below.

▼ **Figure 2.5 Typical AC charging equipment**

2.2.4 Mode 4 charging

In Mode 4 charging, connection of the EV to the AC supply network utilizes an off-board charger where the control pilot function extends to equipment permanently connected to the AC supply. Refer to Figure 2.6 below.

▼ **Figure 2.6 Mode 4 charging**

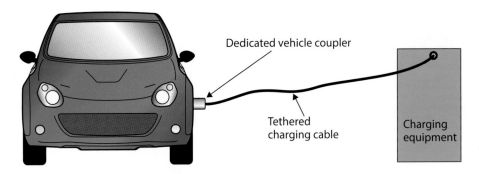

In Mode 4 charging, either single-phase or three-phase AC is converted to DC within the electric vehicle charging equipment. The resulting DC is supplied to the electric vehicle via a charging cable that is tethered to the electric vehicle charging equipment.

Due to its greater complexity, DC charging equipment tends to be in the form of larger units that are usually designed to be wall- or floor-mounted. Refer to Figure 2.7 below.

▼ **Figure 2.7 Typical DC charging equipment**

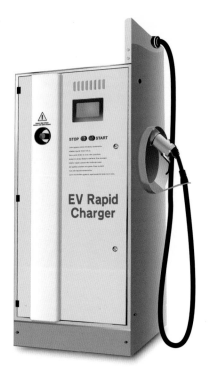

2.3 Socket-outlets, connectors and cables (Regulation 722.55.101)

2.3.1 Modes 1 and 2

Each AC charging point shall incorporate:

(a) one socket-outlet that complies with BS 1363-2 and where the manufacturer approves its suitability for use; or
(b) one socket-outlet or connector that complies with BS EN 60309-2, which is interlocked and is classified in accordance with 6.1.5 of BS EN 60309-1 to prevent the socket contacts being live when accessible; or
(c) one socket-outlet or connector that complies with BS EN 60309-2, is part of an interlocked self-contained product that complies with BS EN 60309-4 and is classified in accordance with 6.1.101 and 6.1.102 to prevent the socket contacts being live when accessible.

2.3.2 Mode 3

Each charging point shall incorporate:

(a) one Type 1 vehicle connector that complies with BS EN 62196-2; or
(b) one Type 2 socket-outlet or vehicle connector that complies with BS EN 62196-2; or
(c) one Type 3 socket-outlet or vehicle connector that complies with BS EN 62196-2.

Note: Vehicle manufacturers' instructions should be followed when determining the type of socket-outlet to be installed.

See Annex A for descriptions and drawings of Types 1, 2 and 3 equipment.

In Mode 3, an electrical or mechanical system shall be provided to prevent the plugging/unplugging of the plug unless the supply has been switched off. Shuttered socket-outlets are not required in Mode 3, as the electrical interlocking that is inherent in Mode 3 provides equivalent protection. The same equivalent protection is provided for vehicle connectors on charging equipment with tethered cables.

2.3.3 Mode 4

Charging equipment designed for Mode 4 charging will always be fitted with a tethered charge cable that is equipped with a dedicated vehicle connector. An electrical or mechanical system shall be provided to prevent the plugging/unplugging of the vehicle connector unless switched off from the supply (Regulation 722.55.101.4).

More detailed information on the charging socket-outlets, plugs, vehicle connectors, vehicle inlets and charging cables is given in Annex A.

2.3.4 Socket-outlet enclosures (Regulation 722.55.101.0.201.2)

BS 7671 requires that each socket-outlet shall be installed in a distribution board in accordance with Regulation 722.51 or in a box in accordance with BS 4662 or BS EN 60670-1 (for example, a flush- or surface-mounted socket-outlet box) and mounted in a fixed position. Portable socket-outlets are not permitted, but tethered vehicle connectors may be used.

2.4 Other variations in electric vehicle charging equipment design and specification

In addition to the types of electric vehicle charging equipment described above, further variations in the design and specification of such equipment can be found. Typical features are described in Sections 2.4.1 to 2.4.6 below. Full details of such features and their operation are not covered by this Code of Practice, but can be found in the installation/operational instructions supplied with the charging equipment being installed.

2.4.1 Power supply

It is important to check that the electrical installation has the appropriate supply arrangement for the EVSE. Some electric vehicle charging equipment is designed to operate from a single-phase AC power supply, whereas other electric vehicle charging equipment may require a three-phase AC power supply.

2.4.2 Multiple socket-outlets

Does the electric vehicle charging equipment meet the needs of the vehicles that will use it? Some types of electric vehicle charging equipment provide multiple socket-outlets. These can either be multiple socket-outlets of the same type or different socket-outlet types to suit different modes of charging, such as one BS 1363-2:2016 socket-outlet for Mode 1 or Mode 2 charging and one BS EN 62196-2:2017 socket-outlet for Mode 3 charging.

2.4.3 Metering

Tariff metering must be provided where use of the charge point is billed on an energy usage basis, and tariff or non-tariff metering may be required for energy management purposes, for example, where Part L2 of the Building Regulations apply.

Some types of electric vehicle charging equipment require a separate feeder pillar containing the energy meter, depending on their installation location. Other types of electric vehicle charging equipment have a built-in energy meter to measure, record and display the amount of electrical energy used.

2.4.4 Protective devices

Some types of electric vehicle charging equipment incorporate either a circuit-breaker and a separate residual-current circuit-breaker (RCCB) or a residual current circuit-breaker with overcurrent protection (RCBO).

Simpler charging equipment such as home chargers may rely on such protection in the consumer or distribution unit that also protects the supply circuit to the charger.

BS 7671:2018 explicitly allows the protection to be split between the charger and the consumer unit – such as DC fault protection in the charger and a Type A residual current device (RCD) in the consumer unit.

2.4.5 Security features

Is the electric vehicle charging equipment appropriate for the installation location in terms of security? Some types of electric vehicle charging equipment, especially those intended for installation in publicly accessible locations, incorporate security features to prevent unauthorized use of the equipment. For instance, the socket-outlet may be located behind a locked cover that can only be unlocked by a registered user presenting their individually authorized electronic key, for example, a radio frequency identification device (RFID).

2.4.6 Communications features

Does the electric vehicle charging equipment provide appropriate remote management, reporting, control and monitoring to meet the user's needs? Some types of electric vehicle charging equipment have built-in communications features that allow data on charge station utilization, energy usage and faults to be communicated to the owner/ operator of the charging equipment. Such data may be communicated either through a wired connection, such as Ethernet, or through a wireless connection, such as General Packet Radio Service (GPRS).

Arrangements prior to installation commencement

The following actions, where applicable, shall be undertaken by a competent person (the designer) prior to the commencement of the electric vehicle charging equipment installation.

Note: In this section, the terms 'designer' and 'installer' are used to align with BS 7671. In some circumstances, the designer and the installer may be the same individual or organization.

3.1 Supply metering

Check whether the incoming supply is a metered supply and establish whether or not it is a landlord's supply.

3.1.1 Unmetered supply

If a public network supply is in use, for example, from a distribution network operator (DNO) or a building network operator (BNO) and it is suspected that this is unmetered, the client should be advised to contact the network operator (DNO or BNO) to arrange for a new connection.

3.1.2 Landlord's supply

If connection to a landlord's supply is being proposed, confirm that the landlord has given permission for such a connection. If this permission has not been given, discuss this with the client. If the landlord refuses connection to their supply, advise the client that it is necessary to contact the DNO or electricity supplier to arrange for a new connection.

3.2 Adequacy of supply

It is the responsibility of the designer to determine that the maximum demand of the existing supply will not be exceeded with the installation of the additional electric vehicle charging equipment.

The designer will first need to assess the capacity of the existing supply. If no records are available, then the designer should contact the DNO for advice, recognizing that they will not be able to determine the rating conclusively unless the supply fuse holder is withdrawn by the relevant DNO, which is likely to incur an additional cost.

Unless design information for an existing installation is available, the installer will need to determine the loads installed, taking diversity into consideration where appropriate. The following IET publications include content related to maximum demand and diversity for installations of varying complexity:

(a)   *Electrical Installation Design Guide: Calculations for Electricians and Designers;* and

(b)   *On-Site Guide*

© The Institution of Engineering and Technology

When considering the additional load to be used for the electric vehicle charging equipment, no diversity shall be applied. Therefore, the rating of the equipment will need to be used in assessing the additional load. For a larger installation with a series of electric vehicle charging equipment connecting points, the assumed maximum load will need to be the sum of all the electric vehicle charging equipment ratings unless there is a load management arrangement limiting the load to a particular value.

Any areas of concern identified whilst assessing the adequacy of the supply shall be discussed with the client. Potential solutions include the following:

(a)   Limiting the maximum current capacity of the charging equipment
This can be achieved by installing charging equipment with a lower capacity, for example 16 A instead of 32 A. Alternatively, the maximum current capacity of some types of electric vehicle charging equipment can be varied between preset levels. For information on whether the charging equipment incorporates this facility and details of how to vary the maximum capacity, refer to the installation/operational instructions supplied with the charging equipment being installed.

(b)   Implementing load management strategies
These could entail electrical, electronic, mechanical or behavioural means of restricting the simultaneous use of electrical equipment with high current demand.

(c)   Uprating the incoming power supply
If this route is chosen, the client shall be advised that it is necessary to contact the DNO or electricity supplier to arrange for an uprating of the incoming power supply.

Even where the supply capacity at individual premises is not exceeded, clusters of connecting points at different premises can have impacts on the local distribution network (i.e. voltage levels and network capacity limits). Examples of planned installation projects with multiple connecting points could be:

(a)   as part of a housing refurbishment programme in the same road; or
(b)   as part of a new housing development.

3.3 Existing earthing arrangements

Check the existing earthing arrangement of the incoming power supply. This information is required to determine what type of earthing arrangement will need to be provided for the electric vehicle charging equipment (refer to Sections 5, 6, 7 and 8).

3.4 Simultaneous contact assessment

If it is planned to provide a separate TT earthing arrangement for the electric vehicle charging equipment and the installation location is in the vicinity of an installation using a different earthing system, for example a network-provided earthing arrangement from a TN-C-S system, carry out an assessment for possible contact between the charging equipment/vehicle being charged and any exposed-conductive-parts or extraneous-conductive-parts connected to the other earthing system, as detailed in Sections 5.3.3, 6.4, 6.7, 7.3 and 8.2.

3.5 GPRS coverage

If the electric vehicle charging equipment is intended to provide communications via GPRS, assess the proposed installation location for available GPRS coverage. Any potential issues identified during this assessment shall be discussed with the client.

3.6 Charging equipment manufacturer's instructions and requirements

Review the specification and any installation instructions and/or requirements provided by the electric vehicle charging equipment manufacturer for the specific piece of equipment being installed. In particular, the installer should take note of:

(a)   whether the electric vehicle charging equipment is fitted with a circuit-breaker and a separate RCCB or an RCBO and the precise specification for any such devices. This information is required by the installer to determine the correct specification of any such devices that may be required for the installation (refer to Sections 5.6, 6.12, 7.5 and 8.5).

(b)   any particular installation requirements that are specific to the type of equipment being installed.

3.7 Planning permission

If planning permission is required for the electric vehicle charging equipment installation, ensure that this has been applied for and granted.

3.8 Traffic Management Order

For on-street, publicly available electric vehicle charging equipment installation, ensure that, where required, any necessary Traffic Management Order (TMO) has been created and duly consulted on.

3.9 Agree installation details with client

Discuss and agree all details of the installation with the client, including:

- the adequacy of the existing electrical installation; see Regulation 132.16 of BS 7671:2018;
- the location of the charging equipment itself;
- the location of any necessary feeder pillar;
- the location of any switches;
- the wiring system, including the routing of all cabling;
- the location of any necessary earth electrode; and
- where appropriate, the location of any circuit-breakers and RCCBs or RCBOs.

SECTION 4

Physical installation requirements

In addition to any installation instructions provided by the charging equipment manufacturer, the physical location and installation requirements covered by this section shall also be followed.

4.1 Potentially explosive atmospheres

4.1.1 General

Electric vehicle charging equipment shall not be installed in locations where a potentially explosive atmosphere exists, for example where petrol vapour or other flammable/combustible gases may be present. Where charging equipment is to be installed in the vicinity of a potentially explosive atmosphere, the electric vehicle charging equipment must be located outside the defined hazardous zone and positioned such that any electric vehicle connected to the charging equipment will also be outside the hazardous zone.

4.1.2 Fuel filling stations

The publication *'Design, construction, modification, maintenance and decommissioning of filling stations'* of the Association for Petroleum and Explosive Administration (APEA) and the Energy Institute (EI) advises:

9.5.11 Electric Vehicle Charging Equipment

Where electric vehicle (EV) charging equipment is installed at filling stations reference shall be made to the latest edition of the IET *Code of Practice for Electric Vehicle Charging Equipment*.

The installation of EV charging equipment at filling stations shall comply with the following criteria:

- The charger and vehicle when charging at the full extent of the charging cable must be outside of any hazardous area.
- When calculating the capacity of the supply required (whether from the filling station supply or a separate Utility Company supply) the full load of all chargers (if more than one) must be allowed – no diversity may be applied.
- BS 7671 Section 722 requires each charge-point to be individually protected by a 30 mA RCD. Therefore, if more than one charge point is to be supplied RCD protection must be provided to achieve selectivity (time and sensitivity) with the 30 mA RCDs.
- If a separate Utility Company supply is provided for the EV charger, the Utility Company supply cable must be routed across or round the filling station forecourt so that it does not pass under any hazardous area.
- If a separate Utility Company supply is provided for the EV charger, it should have a TT earthing system which must be connected to the filling station TT earthing system. It is not acceptable for the EV Charger installation to be connected to an earthing system separate from that of the filling station.
- Where a separate Utility Company supply is provided a prominent Warning label must be mounted on the supply cubicle to indicate that the charger is fed from this separate supply and is not controlled by the PFS main switch.
- Whatever means is adopted to provide the supply to the charger, the supply must be interlocked with the PFS controls so that when the forecourt is closed for any reason, the supply to the charger is also switched off. Where the charging equipment is fed from a supply separate to that serving the filling station (see EN 50174-3), consideration should be given to providing a fibre optic cable link in order to avoid a copper link.
- The EV charger supply must be controlled by the forecourt emergency switching system. Where the charging equipment is fed from a supply separate to that serving the filling station (see EN 50174-3), consideration should be given to providing a fibre optic cable link in order to avoid a copper link.

See also Section 4.10 of the publication *Design, construction, modification, maintenance and decommissioning*.

9.5.2 Battery charging equipment

Battery charging equipment, other than that integral with the dispenser should not be installed or used within any hazardous area of the filling station. Where provision is made for charging batteries integral with electrically energised vehicles, the charging equipment should be located so that the cable connection to the vehicle is within the non-hazardous area when the cable is fully extended.

Where the fuel filling station supply is used, its maximum demand shall be re-evaluated considering the full load of charging equipment (without diversity).

Where EVSE is installed within the curtilage of the fuel filling station:

(a) circuits supplying EVSE shall be controlled by the forecourt emergency switching system;

(b) the EVSE must be interlocked with the filling station controls for lighting and signage, so that power is removed from EVSE when the forecourt is closed;

(c) only TT earthing systems shall be used; and

(d) only EVSE with tethered cabling shall be used in the hazardous area of the filling station, and both the EVSE and parking spaces shall be located so that they cannot be used to charge vehicles in the hazardous area.

Where a separate utility supply is used for EVSE:

(a) a prominent label shall be provided on the EVSE to indicate that it is not controlled by the filling station main switch; and

(b) see Section 5.1.2 of this Code of Practice with regard to simultaneously accessible extraneous- and exposed-conductive-parts and extraneous-conductive-parts of different electrical installations.

Note: If an existing filling station has a protective multiple earthing (PME) supply to forecourt equipment, that may not be compatible with either supplying EVSE, or bonding to a separate TT earth used for EVSE.

It is strongly recommended that contractors working in fuel filling station environments ensure their operatives are trained and qualified in accordance with the Petrol Retail Forecourt Contractors safety passport scheme.

4.2 Location of charging equipment relative to parking space

The electric vehicle charging equipment shall be installed in a position relative to the parking space that ensures that the distance between the charging equipment and the charge connection point on the vehicle, known as the vehicle inlet, is kept to a minimum. This requirement is intended to minimize trip hazards, prevent the charge cable from being strained and helps avoid the temptation to use extension leads which would exacerbate trip hazards.

The Equality Act 2010 applies; accordingly, BS 8300-1 and BS 8300-2, should be considered. Publicly accessible EV charge points should be installed such that they cater for the wide diversity of people who may use them, including wheel-chair users, people with mobility issues and parents with young children.

Note: The use of extension leads for electric vehicle charging is not permitted.

4.2.1 Charging equipment intended to be used for charging only one type of vehicle

For charging equipment intended to be used for charging only one type of vehicle, for example an industrial installation used for charging a fleet of identical vehicles, the location of the charging equipment relative to the parking space shall be assessed using the relevant vehicle type.

4.2.2 Charging equipment intended to be used for charging multiple types of vehicle

For charging equipment intended to be used for charging a variety of different vehicle types, or where the type of vehicle to be charged is unknown, compliance with the requirement specified in Section 4.2 is best achieved by installing the charging equipment at one corner of the parking space. This is because the position of the vehicle inlet on the vehicle is not currently standardized (some vehicles have their inlet on the front of the vehicle, some on the left-hand side of the vehicle and some on the right-hand side).

Figure 4.1 on the following page demonstrates how placing the charging equipment at one corner of the parking space(s) can facilitate charging both a vehicle with its vehicle inlet on the front of the vehicle (labelled as A) and a vehicle with its vehicle inlet on the rear right-hand side of the vehicle (labelled as B).

▼ Figure 4.1 Charging equipment location relative to parking space

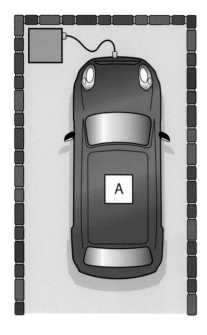 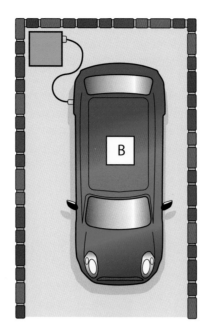

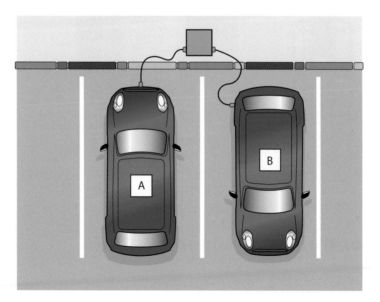

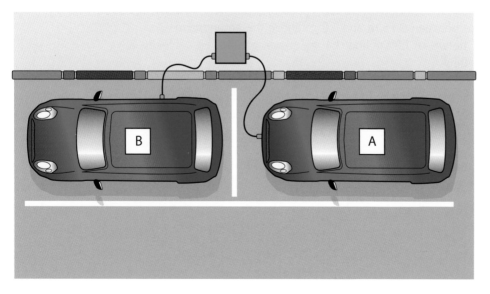

4.3 Protection against vehicle impact

The electric vehicle charging equipment shall be installed in a position to minimize the likelihood of vehicle impact damage, for example to avoid the possibility of vehicles inadvertently driving into or reversing into the equipment.

Where it is not possible to install the electric vehicle charging equipment in a position that minimizes the risk of vehicle impact, additional protective barriers shall be installed (see Figure 4.2). Where electric vehicle charging equipment is associated with an accessible parking space, clear access should be provided for a wheelchair user to operate the equipment, including ensuring that access to socket- outlets and controls is not impeded by bollards, as required by BS 8300-1:2018.

▼ **Figure 4.2 Typical protective barriers**

4.4 Location of controls and socket-outlets

It is recommended that the electric vehicle charging equipment be installed such that:

(a) the main operating controls (including, where applicable, payment means) and any socket-outlets are between 0.75 m and 1.2 m above the ground; and
(b) display screens are between 1.2 m and 1.4 m above ground, viewable by a person standing or sitting.

This is to avoid damage from vehicle bumpers and to ensure their accessibility in accordance with the British Standards for inclusive built environments, BS 8300-1:2018 and BS 8300-2:2018. Refer to Figure 4.3.

Note: Regulation 722.55.101.5 also contains standards for the height of socket-outlets for electrical safety. The recommended band for inclusive use is compliant with BS 7671.

▼ **Figure 4.3 Height of controls and socket-outlets**

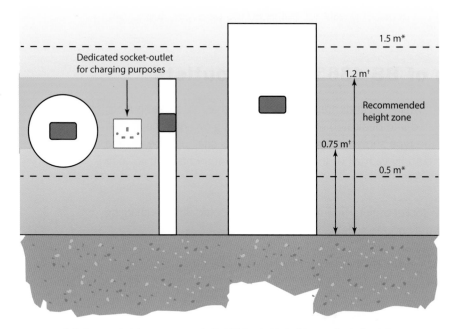

* Maximum and minimum recommended heights for electrical safety according to Regulation 722.55.101.5 of BS 7671:2018.
† Maximum and minimum recommended heights of socket-outlets and controls in BS 8300-1 and BS 8300-2. Displays are recommended to be within the height range 1.2 m to 1.4 m.

4.5 Free space around the charging equipment

Once installed, and with any protective barriers in place, there must be sufficient space around the electric vehicle charging equipment to allow the full opening of any doors or covers provided for operational use, inspection or maintenance of the charging equipment. The installer shall refer to the installation or operational instructions supplied with the charging equipment being installed for details.

4.6 Ventilation and cooling

Once the electric vehicle charging equipment is installed, there must be sufficient space around it to allow for adequate ventilation and cooling of the equipment. This is particularly important for DC charging equipment incorporating rectifiers. The installer shall refer to the installation or operational instructions supplied with the charging equipment being installed for details.

4.7 Avoidance of trip hazards

The electric vehicle charging equipment shall be installed in such a way as to avoid creating unnecessary trip hazards.

Any electrical wiring connecting the charging equipment to the mains supply shall be routed to avoid creating potential trip hazards, and suitably clipped or enclosed in containment.

Provision shall be made for the safe storage of any tethered charging cables when they are not in use.

4.8 Avoidance of unnecessary obstruction

The electric vehicle charging equipment shall not be installed in a position that causes an unnecessary obstruction to public or private footpaths, access passages, and so on.

4.9 Labelling of BS 1363 socket-outlets

Any BS 1363 socket-outlets installed as dedicated electric vehicle charging socket-outlets, but which are not incorporated into dedicated electric vehicle charging equipment, shall be identified by a label of the type shown in Figure 4.4. This label shall only be applied to socket-outlets that fully comply with all of the applicable requirements of this Code of Practice.

This labelling is to identify socket-outlets that are suitable for electric vehicle charging and, by default, those that may not be suitable.

▼ **Figure 4.4 Label for socket-outlets suitable for electric vehicle charging**

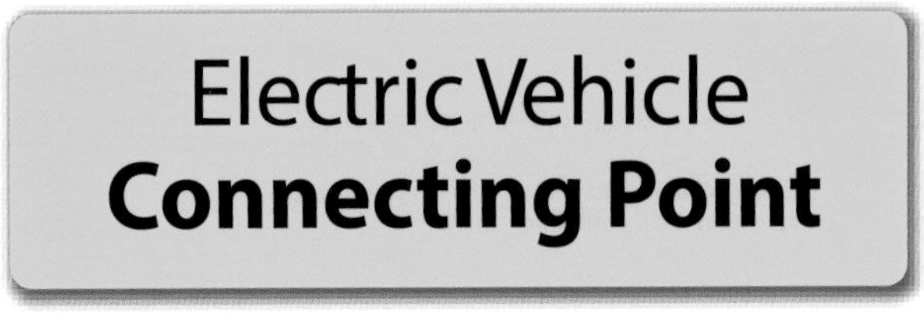

Electric Vehicle **Connecting Point**

SECTION 5

Electrical requirements – General

This section contains general requirements for the electrical installation supplying electric vehicle charging equipment and is applicable to all locations. For specific requirements applicable to domestic installations, refer to Section 6; for on-street installations, refer to Section 7 and for commercial and industrial installations, refer to Section 8.

5.1 General

5.1.1 BS EN 61851 compliance

The charging equipment shall comply with BS EN 61851 *Electric vehicle conductive charging system* including:

Part 1: General requirements

Part 21: Electric vehicle requirements for conductive connection to an AC/DC supply

Part 22: AC electric vehicle charging stations

Part 23: DC electric vehicle charging stations

Part 24: Digital communication between a DC electric vehicle charging station and an electric vehicle for control of DC charging

5.1.2 BS 7671 compliance

The charging equipment electrical installation shall comply with BS 7671 *Requirements for Electrical Installations*, including the particular requirements of Section 722: Electric Vehicle Charging Installations.

Section 722 of BS 7671 modifies or prohibits certain of the General Requirements of Parts 4 and 5 of that Standard.

Since electric vehicle charging installations often supply equipment away from existing buildings, the following key requirements of BS 7671 in respect of protective earthing and bonding should be noted, regardless of what earthing arrangement is chosen for the charging installation:

(a) Regulation 411.3.1.1 requires simultaneously accessible exposed-conductive-parts to be connected to the same earthing system. This applies whatever the earthing arrangement.
(b) The requirements of Chapter 54 for earthing and bonding (including main bonding), in particular, in relation to extraneous-conductive-parts that may also be extraneous-conductive-parts of other earthing systems.
(c) Regulation 542.1.3.3 in relation to protective conductors common to electrical installations with separate earthing arrangements. This may include screens or 'ground' conductors in communication and control cables.

5.2 Protection against electric shock

5.2.1 General requirements

The following protective measures shall not be used for electric vehicle charging installations:

(a) obstacles and placing out of reach (Regulation 722.410.3.5); and
(b) earth-free local equipotential bonding (Regulation 722.410.3.6).

5.2.2 TN-C systems

The final circuit supplying a charging point shall not include a PEN (combined neutral and protective) conductor. (See Regulation 722.312.2.1).

Note: In Great Britain, Regulation 8(4) of the ESQCR 2002 prohibits the use of PEN conductors in consumers' installations.

5.2.3 TN-S systems

Unless it can be guaranteed by the relevant DNO that the TN-S supply system is TN-S back to source and will not be converted to TN-C-S as part of any ongoing network upgrades, such installations shall be treated as TN-C-S systems (PME supply).

For electric vehicle charging equipment installations that are part of a guaranteed TN-S system, for example where the charging equipment is supplied from a TN-S installation supplied by its own 11 kV transformer, the charging equipment may be connected to the existing earthing arrangements, whether the charging equipment is installed within a building or not. The installer must confirm that the earthing and bonding arrangements meet the current requirements of BS 7671 for TN-S systems, and that any non-compliances are rectified.

5.2.4 TN-C-S systems

Regulation 722.411.4.1 imposes particular requirements for electric vehicle charging installations where they are a part of a TN-C-S system and a PME earthing facility is used. The requirements are copied below as they are significant to the installation design.

> **722.411.4.1** A PME earthing facility shall not be used as the means of earthing for the protective conductor contact of a charging point located outdoors or that might reasonably be expected to be used to charge a vehicle located outdoors unless one of the following methods is used:
>
> **(i)** The charging point forms part of a three-phase installation that also supplies loads other than for electric vehicle charging and because of the characteristics of the load of the installation, the maximum voltage between the main earthing terminal of the installation and Earth in the event of an open-circuit fault in the PEN conductor of the low voltage network supplying the installation does not exceed 70 V rms.
>
> > **NOTE 1:** Annex 722, item A722.2 gives some information relating to (i).
> > **NOTE 2:** See also Regulation 641.5 when undertaking alterations and additions.
>
> **(ii)** The main earthing terminal of the installation is connected to an installation earth electrode by a protective conductor complying with Regulation 544.1.1. The resistance of the earth electrode to Earth shall be such that the maximum voltage between the main earthing terminal of the installation and Earth in the event of an open-circuit fault in the PEN conductor of the low voltage network supplying the installation does not exceed 70 V rms.
>
> **(iii)** Protection against electric shock is provided by a device which disconnects the charging point from the live conductors of the supply and from protective earth in accordance with Regulation 543.3.3.101(ii) within 5 s in the event of the voltage between the circuit protective conductor and Earth exceeding 70 V rms. The device shall not operate if the voltage exceeds 70 V rms for less than 4 s. The device shall provide isolation. Closing or resetting of the device shall be by manual means only, Equivalent functionality could be included within the charging equipment.
>
> Where buried in the ground, a protective conductor connecting to an earth electrode for the purposes of (ii) or (iii) shall have a cross-sectional area not less than that stated in Table 54.1.

5.2.5 Street electrical fixtures

On-street installations are discussed in Section 7.

5.3 PME supply

5.3.1 Introduction

Most premises in the UK have a PME supply. This has proven to be a very reliable method of providing electrical installations with an earth connection. However, particular precautions are required for the earthing of installations and these are specified in BS 7671:2018, in particular, in Chapter 54. In the event of an open-circuit fault to the neutral (PEN conductor) of a TN-C-S (PME) system:

- all conductive-parts connected to the earthing system can rise to a dangerous voltage above earth potential;
- load currents may return to the origin of the supply via the protective conductors of the installation, which can cause overheating of the earthing and bonding conductors.

This is illustrated in see Figure 5.1 below.

The installation of an RCD will not provide any protection against these effects.

▼ **Figure 5.1 Schematic of a TN-C-S (PME) system with an open-circuit PEN conductor**

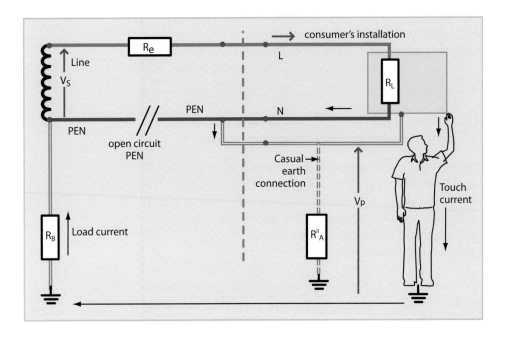

To prevent anyone within the building being subjected to this voltage, protective equipotential bonding of all extraneous-conductive-parts is required by BS 7671. This may include as required by Regulation 411.3.2:

- water installation pipes
- gas installation pipes
- other installation pipework and ducting;
- central heating and air-conditioning systems; and
- exposed metallic structural parts of the building.

The bonding conductors must be of sufficient cross-sectional area to carry the load current that would flow in the event of a fault in the electricity supply. The required cross-sectional areas are given in Table 54.8 of BS 7671:2018, which is reproduced in Table 5.1. The minimum cross-sectional area is 10 mm^2.

Note: Local distributors' network conditions may require a larger conductor.
Note: The neutral of supply refers to the neutral of the supplier's feeder cable at the supplier's cut-out, and not the tails between the supplier's cut-out and meter, or from the meter to the consumer's equipment.

▼ Table 5.1 Minimum cross-sectional area of the main protective bonding
conductor in relation to the neutral of the supply
(from Table 54.8 BS 7671:2018)

Copper equivalent cross-sectional area of the supply neutral conductor	Minimum copper equivalent* cross-sectional area of the main protective bonding conductor (mm²)
35 mm² or less	10
over 35 mm² up to 50 mm²	16
over 50 mm² up to 95 mm²	25
over 95 mm² up to 150 mm²	35
over 150 mm²	50
*The minimum copper equivalent cross-sectional area is given by a copper bonding conductor of the tabulated cross-sectional area or a bonding conductor of another metal affording equivalent conductance.	

This bonding will not protect people outside the building who are simultaneously in contact with conductive-parts connected to the PME earth (for example an electric vehicle being charged) and true Earth (for example metalwork in contact with the ground, but not connected to the main earthing terminal of the building).

5.3.2 Precautions to be taken for PME supplies

BS 7671 does not allow a PME earthing facility to be used to supply an electric vehicle charging point unless the requirements of Regulation 722.411.4.1 are met. These requirements are considered in Section 6 for dwellings, Section 7 for on-street installations and Section 8 for commercial and industrial installations.

The alternative is to provide protection against electric shock by measures that do not make use of the PME earth, including:

(a) automatic disconnection using a TT system (refer to Section 5.3.3);
(b) automatic disconnection using an isolating transformer to separate the earths (refer to Section 5.3.4); or
(c) adoption of the protective measure of electrical separation (refer to Section 5.3.6).

5.3.3 Converting to a TT system

There are two options:

(a) make the circuit that supplies the charging equipment part of a TT system.

Note: In cases where precautions must be taken to prevent simultaneous contact between those exposed-conductive-parts or extraneous-conductive-parts connected to the TT earthing system and exposed-conductive-parts or extraneous-conductive-parts connected to the TN-C-S (PME) earth of the main installation, a risk assessment is required – see Annex B, C, D or E as relevant.

(b) convert the whole installation to a TT earthing system.

Note: These options should not be adopted where there would be simultaneous access to conductive parts of an adjacent installation that is connected to a PME earthing arrangement. A risk assessment is required – see Annex B, C, D or E as relevant.

When assessing whether to adopt TT in an existing installation, consideration must also be given to the safety of installing earth electrodes, and the risk of striking underground services. For some installations, it may not be possible to provide suitable earth electrodes safely, without disruptive excavation.

When all of the above points are considered, converting part or all of some existing installations to TT may not be practicable.

5.3.3.1 Providing a TT earthing system for the charging equipment

Providing a dedicated TT earthing system for the charging equipment avoids any problems associated with an open-circuit fault in the neutral (PEN conductor) of the supply, as the charging equipment would not be connected to the PME supply earth.

However, providing a dedicated TT earthing system is only acceptable if there is no possibility of simultaneous contact between exposed- and/or extraneous-conductive-parts of the TT earthing system and exposed- and/or extraneous-conductive-parts of the TN-C-S (PME) earthing system of the main installation or any other earthing system of any other nearby installation.

The earth electrode zone of the TT system must not overlap the zone of any extraneous-conductive-parts of the main installation. If it does, the benefit of separating the earthing systems will be lost. A risk assessment is required: see Annex B.

The resistance of the earth electrode shall not exceed 200 Ω and an RCD or RCDs shall be provided to protect the electric vehicle charger installation.

A notice must be fixed at suitable positions, including, but not limited to, the origin, the meter position if remote from the origin, and at the designated point of isolation for the electric vehicle charging equipment supply to advise any person carrying out work in the future that the electric vehicle charging equipment is connected to a separate TT earthing system. See the example in Figure 5.2.

▼ **Figure 5.2 Example of warning notice for TT charging equipment installed in locations using a PME earth for other circuits, or where premises with PME supply is converted to TT**

ELECTRIC VEHICLE CHARGING EQUIPMENT
TT system adopted for charging equipment circuit.
Do not connect to the PME earth

5.3.3.2 Converting the whole installation to a TT earthing system

Converting the whole installation to a TT earthing system avoids problems associated with an open-circuit fault in the neutral (PEN conductor) of the supply, as the complete installation is not to be connected to the PME earthing arrangement. Additionally, this will mean that the electric vehicle charging equipment (including the connecting point) and the vehicle on charge need not be situated so as to prevent simultaneous contact between these items and exposed- or extraneous-conductive-parts of the main installation.

However, converting the whole installation to TT is usually only suitable for smaller installations, such as dwellings. Converting a larger installation to TT may present difficulties in co-ordinating RCD protection. Issues that can be introduced when converting PME systems to TT include the following:

(a) because earth fault loop impedances increase, overcurrent protective devices cannot be relied upon alone to provide automatic disconnection in the event of an earth fault. Additional protection using RCD is therefore necessary.

(b) in the event of a fault to earth in a TT system, unless both the supply and installation earth electrode resistances are extremely low, most of the voltage due to earth fault current appears in the earth electrode resistances. Therefore, touch voltages between the general mass of earth and the installation protective earthing (including exposed-conductive-parts and extraneous-conductive-parts) may well be greater than with a TN system; however, touch voltages between exposed-conductive-parts, and between exposed-conductive-parts and extraneous-conductive-parts, may be much lower than with a TN system.

(c) there is no longer a low impedance path to the general mass of earth, and between the supply neutral and the installation protective earth. This can affect electromagnetic compatibility, and make the electrical installation (and equipment within it) more susceptible to surges due to switching and of atmospheric origin. Electromagnetic compatibility and surge protection assessments may need to be reviewed.

A risk assessment must be carried out to consider the risks of simultaneous contact with conductive-parts connected to adjacent TN systems (if any).

If a property is converted from PME to TT a notice must be fixed at the cut-out position to advise any person carrying out work that the PME earthing facility is not connected. See the example in Figure 5.2.

5.3.4 Installing an isolating transformer

Installing an isolating transformer in the electric vehicle charging equipment circuit avoids any problems associated with an open-circuit fault in the neutral (PEN conductor) of the supply, as the isolating transformer electrically separates the PME supply earth from the charging equipment. See Figure 5.3. below.

▼ **Figure 5.3 Charging equipment separated from the TN-C-S (PME) supply earth by a transformer**

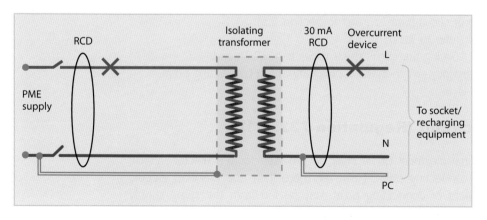

Note: Figure 5.3 shows an isolating transformer installation utilizing automatic disconnection as the method for protection against electric shock, not electrical separation.

Note: As the protective conductor labelled PC will be connected to the vehicle body, and perhaps other parts of the charging equipment, it is entirely feasible that PC may come into (accidental) contact with the general mass of Earth. It is therefore necessary to ensure that there is no possibility of simultaneous contact of persons between PC and the PME supply circuit protective conductor.

The fixed isolating transformer shall comply with BS EN 61558-2-4 and shall be installed such that the secondary circuit protective conductors are separated from the general installation earthing.

An overcurrent device and an RCD shall be installed in the secondary circuit.

5.3.5 Installing equipment that detects open-circuit faults

Installing equipment that detects open-circuit faults in the neutral (PEN conductor) and disconnects the supply, including the protective conductor, in the event of such faults, would address the problems associated with an open-circuit fault in the neutral (PEN conductor). However, although such devices are currently under development, at the time of writing of this Code of Practice, none are commercially available. If they are made available, these devices may be an appropriate solution for single-phase installations.

5.3.6 Electrical separation

Where electrical separation is used to provide protection against electric shock for the electric vehicle charging equipment circuit, all of the requirements of Section 413 of BS 7671:2018 shall be met.

5.4 General requirements for circuit design, cable selection and dimensions

5.4.1 BS 7671 Compliance

The electrical circuit for the electric vehicle charging equipment shall comply with the requirements of BS 7671 and shall include the following:

- a suitably rated fuse or circuit-breaker (refer to Section 5.5);
- a suitably rated RCD (refer to Section 5.6);
- a suitable means of isolating the circuit (refer to Section 5.7.2); and
- if necessary, an emergency switch in close proximity to the connecting point (refer to Section 5.7.1).

5.4.2 Separate circuit (Regulation 722.311)

A separate, dedicated final electrical circuit shall be provided for the electric vehicle charging equipment or connecting point. However, more than one piece of electric vehicle charging equipment or connecting point may be fed from the same supply circuit, provided that the combined current demand of the equipment does not exceed the rating of the supply circuit.

5.4.3 Cable selection

The cable types and cross-sectional areas for the electric vehicle charging equipment circuit shall be chosen to ensure compliance with the requirements for:

- protection against electric shock (refer to Section 5.5);
- overcurrent protection (Chapter 43 of BS 7671:2018); and
- voltage drop (Section 525 of BS 7671:2018).

No diversity shall be allowed for final circuits, unless automatic controls limit the demand.

Diversity may be allowed for a dedicated distribution circuit that supplies multiple electric vehicle charging points if load control is available (Regulation 722.311).

When specifying the cable sizes for use in the electric vehicle charging equipment circuit, reference shall be made to either Appendix 4 of BS 7671, which gives current-carrying capacity and voltage drop for cables, or manufacturer's data.

5.5 Protection against electric shock

As required by BS 7671, basic protection against electric shock shall be provided by automatic disconnection of supply or electrical separation.

5.5.1 Automatic disconnection of supply

Where automatic disconnection of the supply is used to provide protection against electric shock for the electric vehicle charging equipment circuit, compliance with Section 411 of BS 7671 shall be ensured. (See Sections 5.3.2 and 5.3.3.)

5.5.2 Electrical separation

Where electrical separation is used to provide protection against electric shock for the electric vehicle charging equipment circuit, all the requirements of Section 413 of BS 7671 shall be met.

The electric vehicle charging equipment circuit shall be supplied through a fixed isolating transformer complying with BS EN 61558-2-4 and live parts of the separated circuit shall not be connected at any point to another circuit, to earth or to a protective conductor.

This protective measure shall be limited to the supply of one electric vehicle supplied from one unearthed source with simple separation. (See Section 5.3.6).

Note: This is not the arrangement shown in Figure 5.3.

5.5.3 Additional protection

Additional protection shall be provided by the installation of RCDs as specified in Section 5.6.

5.6 Requirements for the provision of RCDs

5.6.1 RCD selection and installation

Each connecting point shall be protected by a 30 mA RCD, with an operating time not exceeding 40 ms at a residual current of 5 $I_{\Delta n}$, appropriately selected for the nature of the residual fault currents expected.

BS 7671 has specific requirements for the RCDs, as follows:

> **722.531.2 RCDs**
>
> 722.531.2.101 Except for circuits using the protective measure of electrical separation, each charging point shall be protected by its own RCD of at least Type A, having a rated residual operating current not exceeding 30 mA.
>
> Each charge point incorporating a socket-outlet or vehicle connector complying with the BS EN 62196 series, protective measures against DC fault current shall be taken, except where provided by the EV charging equipment. The appropriate measures, for each connection point, shall be as follows:
>
> - RCD type B; or
> - RCD type A and appropriate equipment that ensures disconnection of the supply in case of DC fault current above 6 mA.
>
> RCDs shall comply with one of the following standards: BS EN 61008-1, BS EN 61009-1, BS EN 60947-2 or BS EN 62423.
>
> **Note:** Requirements for the selection and erection of RCDs in the case of supplies using DC vehicle connectors according to the BS EN 62196 series are under consideration.
>
> 722.531.2.1.1 RCDs shall disconnect all live conductors.

5.6.2 Discrimination (selectivity)

There should be discrimination (selectivity) between any RCD installed at the connecting point or incorporated in the charging equipment and the protection at the origin of the circuit.

Under some circumstances, due to other requirements of BS 7671, discrimination is not practicable, for example, where part of a circuit supplying the connecting point is concealed in a wall at a depth less than 50 mm.

5.6.3 RCD types and standards

Only RCDs complying with BS EN 61008, BS EN 61009, BS EN 62423 or equivalent, that disconnect all live conductors, including the neutral, shall be used. BS 7671 requires an RCD that is at least Type A. Type AC are not suitable.

A Type A RCD is an RCD for which tripping is ensured for residual sinusoidal alternating currents; residual pulsating direct currents; and residual pulsating direct currents superimposed by a smooth direct current of 6 mA, with or without phase-angle control, independent of the polarity.

A Type B RCD is an RCD for which tripping is ensured as for: Type A for residual sinusoidal currents up to 1 kHz; residual sinusoidal currents superimposed by a pure direct current; pulsating direct currents superimposed by a pure direct current; and residual currents which may result from rectifying circuits, three-pulse star connection or six-pulse bridge connection, two-pulse bridge connection line-to-line with or without phase-angle monitoring, independently of the polarity.

Note: Single-module RCBOs designed to fit an existing consumer unit or distribution board may not comply with this requirement, as the neutral may be a solid link that does not disconnect upon tripping. Some single-phase RCBOs may disconnect all poles, and therefore be suitable, but may only incorporate overcurrent detection in the phase, not the neutral.

5.6.4 Labels

A label describing the operation and testing of the RCD, as shown in Figure 5.4, shall be fitted in a visible position, adjacent to the installed location of each device.

▼ **Figure 5.4 Label for the regular cycling of an RCD**

> This installation, or part of it, is protected by a device which automatically switches off the power supply if an earth fault develops. **Test every six months** by pressing the button marked **'T'** or **'Test'.** The device should switch off the supply and should then be switched on to restore the supply. If the device does not switch off the supply when the button is pressed seek expert advice.

5.7 Requirements for isolation and switching

5.7.1 Emergency switch

Because all electric vehicle charging equipment and connecting points are required to be protected by a 30 mA RCD, an emergency switch is not usually required. However, the particular needs of the installation and the charging equipment manufacturer's instructions shall be considered. If an emergency switch is provided, it shall be located in a position that is readily accessible and shall be suitably identified by marking and/ or labelling.

5.7.2 Isolation

A means of isolating the electric vehicle charging equipment circuit shall be provided, in accordance with Regulation 537.2 of BS 7671:2018. This isolating device shall be located in a position that is readily accessible for maintenance purposes and shall be suitably identified by marking and/or labelling.

5.7.3 Functional switching

Additional forms of switching for functional purposes may be required, for example, to prevent the unauthorized use of socket-outlets located outside buildings that may be accessible to the public. These shall be discussed and agreed with the client.

5.8 IP ratings

5.8.1 IP rating of charging equipment (Regulation 722.512.2)

The electric vehicle charging equipment being installed shall have an IP rating suitable for the proposed installation location. For example, charging equipment that is to be installed in an outdoor location shall have an IP rating of at least IP44.

5.8.2 IP rating of installation components

The installer shall ensure that any connectors, glands, conduits, etc. used to connect the power supply to the electric vehicle charging equipment are of an appropriate IP rating, and are installed in such a way as to maintain a suitable IP rating, in accordance with BS 7671.

5.8.3 Impact

Equipment shall be protected against foreseeable damage by position and/or by external protection.

Electric vehicle charging equipment installed in publicly accessible locations, for example, car parks, shall be protected against a minimum impact severity of AG2, as defined in Appendix 5 of BS 7671:2018.

5.9 Socket-outlets and connectors

The requirements are found in Regulation 722.55.101 and discussed in Section 2.3.

5.10 Lightning protection systems

Where a building has a lightning protection system, reference shall be made to BS EN 62305 and Regulation 411.3.1.2 of BS 7671:2018.

5.11 Installations having alternative sources of supply

Some installations may have an alternative source of supply that can be used when the public supply fails. Examples include:

(a) energy storage systems that can operate in island mode (that is, with the public supply disconnected); and
(b) back-up generators/uninterruptible power supplies (UPS).

In such systems, when the alternative source is used:

(a) the means of earthing may change; and
(b) loop impedances will change, and in many systems will reduce, sometimes dramatically.

In relevant installations, the above factors should be taken into account if the electric vehicle charging equipment is to be powered from both the alternative supply and the public supply.

The projected wider adoption of electrical energy storage systems may increase the number of installations that are affected. Section 9.4 of the IET *Code of Practice for Electrical Energy Storage Systems* provides guidance for ensuring electrical safety in island mode operation and for assessing the impact of an alternative means of earthing and impact on earth fault loop impedances.

Electrical requirements – Dwelling installations

6.1 Scope

This section applies to premises that can be classified as a self-contained unit designed to accommodate a single household, or any outbuildings supplied by the electrical installation of this type of premises, such as a detached garage.

The following types of premises are not included:

(a) buildings exclusively containing rooms for residential purposes, such as nursing homes, student accommodation and similar; and

(b) buildings that accommodate more than one household, such as flats.

6.2 Demand

In dwellings, the total demand, including the electric vehicle charging equipment, should not exceed the rating of the supply cut-out after consideration of diversity. Some charging equipment may include demand-limiting capabilities (see Section 12.1).

Combined with time-of-use tariffs, charging controls can allow charging to be carried out at reduced rates.

It is normally unnecessary to incorporate hardware into the installation to achieve timer control of electric vehicle charging as most electric vehicles have such controls built in already.

6.3 Earthing and protective equipotential bonding requirements

The earthing system requirements for the electric vehicle charging equipment will depend upon the earthing arrangements of the electrical supply to the charging equipment and upon the location of the charging equipment and/or the vehicle being charged, i.e. whether the charging equipment and vehicle being charged are within a building or outside.

6.4 TT systems

For electric vehicle charging equipment installations that are to be connected to an existing TT system, the charging equipment may be connected to the existing earthing arrangement whether the charging equipment is installed within a building or not. The installer must confirm that the earthing and bonding arrangements meet the requirements of BS 7671 for a TT system, and that any deficiencies are rectified.

If any adjacent installations have PME earthing arrangements, a risk assessment must be carried out to identify any hazards and any precautions necessary. A copy of the risk assessment shall be included in the installation documents given to the customer – see Annex B.

6.5 TN-S system

Unless it can be guaranteed by the relevant DNO that the TN-S system is TN-S back to source and will not be converted to TN-C-S as part of any ongoing network upgrades, such installations shall be treated as TN-C-S systems (PME supply).

For electric vehicle charging equipment installations that are part of a guaranteed TN-S system, the charging equipment may be connected to the existing earth whether the charging equipment is installed within a building or not. The installer must confirm that the earthing and bonding arrangements meet the requirements of BS 7671 for TN-S systems, and that any non-compliances are rectified.

6.6 TN-C-S system (PME supply)

6.6.1 General

When an electric vehicle charging point is to be supplied from an installation in a dwelling with a PME supply, the earthing and protective bonding arrangements to be adopted require careful consideration.

Regulation 722.411.4.1 of BS 7671:2018 includes the provision that PME (TN-C-S) supplies may not be used for electric vehicle charging supplies, unless one of three conditions are met: – these are numbered (i), (ii) and (iii) in the Regulation. Examining the three conditions:

(i) refers to three-phase installations, so will only be applicable very infrequently for dwellings.
(ii) requires the installation of earth electrodes of a suitably low earth electrode resistance Z_{EE}.
(iii) describes a voltage operated device that, at the time of publication, is not available, but when/if it is available, may be an appropriate solution for dwellings. BS 7671 states that this device may be incorporated into charging equipment itself.

For electric vehicle charging points located outdoors (or that might reasonably be expected to charge vehicles outdoors), the installer may elect one of the following options:

(a) to install (or use charging equipment containing) a voltage operated trip, if available (Regulation 722.411.4.1(iii));
(b) to use the PME earth and install earth electrodes that will reduce the touch voltage in the event of an open circuit PEN in the supply to 70 V;
(c) to make the charging point part of a TT installation (this will also require earth electrodes; see Section 6.8)
(d) to install an isolating transformer (see Section 5.3.4); or
(e) to adopt electrical separation (see Section 5.3.6)

The touch voltage limit of 70 V was adopted after reference to Table 2c of IEC 60479-5 *Effects of current on human beings and livestock - Part 5: Touch voltage threshold values for physiological effects*, Edition 1. This proposes a limit of 71 V for both-hands-to-feet, in water-wet conditions with medium contact area (12.5 cm²).

6.7 Using the PME earth

6.7.1 Indoor charging point

For vehicle connecting points installed such that the vehicle can only be charged within a building, for example, in a garage with a (non-extended) tethered lead, the PME earth may be used without additional earth electrodes (see Figure 2.4).

6.7.2 Outdoor PME charging point

For single-phase installations, a voltage-operated device that will meet 722.411.4.1(iii), when and if available, may be an appropriate solution. (At the time of writing these devices are not available.)

The installer may also adopt the measures discussed in Section 5.3.4 (Installing an isolating transformer) and Section 5.3.6 (Electrical separation).

The installer may also meet the requirements of the regulations when using a PME earth by installing an earth electrode system that will reduce the touch voltage in the event of an open-circuit fault in the PEN conductor of the network supplying the installation, to 70 V rms (Regulation 722.411.4.1 (ii)). See Figure 6.1 below.

▼ **Figure 6.1 – PME installation with additional earth electrode (R$_A$)**

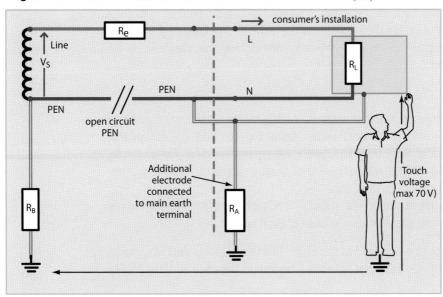

The ways of keeping the touch voltage down are:

(a) ensuring that the electrode resistance to earth (R$_A$) is as low as is reasonably practicable; and

(b) limiting the charging current; and

(c) bonding extraneous-conductive-parts.

6.7.3 Installing an earth electrode system

See Annex G

6.8 TT charging circuits from PME supplies

The options for installing a TT charging point are:

(a) for outdoor connecting points, when a risk assessment shows that the electric vehicle charging equipment, connection point and vehicle on charge are not simultaneously accessible with conductive-parts connected to a PME earthing arrangement (for example, where the separation is more than 10 m), the charging circuit may be part of a TT installation.

Note: Remote garages may already be supplied from a TT arrangement that has been derived from a TN-C-S (PME) supply in the house.

(b) when outdoor-installed electric vehicle charging equipment (and/or the connecting point) is to be installed close to the house, these items may be supplied from a TT supply obtained from the house's TN-C-S (PME) system, provided that:

(i) a risk assessment is undertaken to assess the possibility of simultaneous contact between any accessible conductive-parts connected to a PME earth, and:

- any conductive-parts connected to the TT earth electrode, for example, the electric vehicle charging equipment (including the connecting point);
- the vehicle on charge;

AND

(ii) suitable precautionary measures are put in place to prevent all such identified means of simultaneous contact, for example, by covering a metal pipe with insulating material, inserting one or more lengths of plastic pipe in a metal pipe run, or by replacing a Class I luminaire with a Class II luminaire, etc.

(c) if a risk assessment shows that the risk of simultaneous contact between any accessible conductive-parts connected to a PME earth, and any extraneous- or exposed-conductive-parts associated with outdoor-installed vehicle charging equipment (including the connecting point and the vehicle on charge etc.) cannot be prevented, the adoption of a TT system for the whole installation can be considered (see Section 5.3.3.2).

Notes:

- When assessing the risks of simultaneous contact between conductive-parts in contact with different earthing systems within items (a), (b) and (c) above, this assessment must include all accessible conductive-parts including, for example, any conductive-parts associated with adjoining or nearby properties, and any other outdoor-installed electrical equipment, etc.
- The written risk assessment document must in each case be included with the installation documentation (Installation Certificate, etc.): see Annex B.

A suitable advisory label must be provided at the origin of every circuit supplied forming part of the TT system stating that this must not be connected to the PME earthing facility (see Sections 5.3.3.1 and 5.3.3.2).

Figure 6.2 shows circuit schematics for a typical domestic installation with a PME supply where the electric vehicle charging equipment is installed such that conductive-parts associated with the connecting point and the exposed-conductive-parts of the vehicle are not simultaneously accessible with conductive-parts of the main installation connected to a PME earthing arrangement.

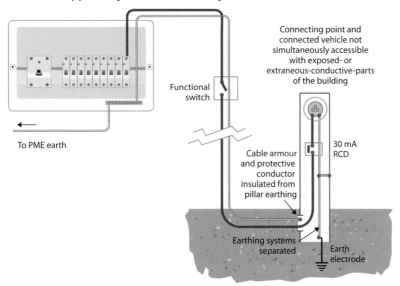

Note: Neutral omitted for clarity.

6.9 Protective equipotential bonding

At the time of installation, it shall be confirmed that the earthing and bonding arrangements comply with the requirements of the current edition of BS 7671.

Additionally, all extraneous-conductive-parts and exposed-conductive-parts accessible from the connecting point or any vehicle being charged shall be connected by supplementary protective bonding conductors. This is particularly important where the vehicle is to be charged in a domestic garage, due to the large expanse of exposed-conductive-part (the vehicle's body) and the restricted space in a garage with the vehicle parked inside it.

Figure 6.3 shows the supplementary bonding for a domestic installation with the connecting point installed in a garage.

▼ Figure 6.3 – Supplementary bonding for a domestic installation with the connecting point installed in a garage

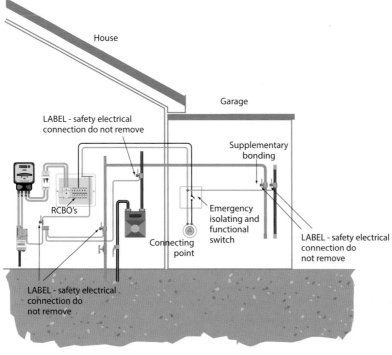

6.10 General requirements for circuit design, cable specifications and dimensions

6.10.1 BS 7671 compliance

The electrical circuit for the electric vehicle charging equipment shall comply with the requirements of BS 7671, in particular, Section 722.

6.10.2 Dedicated final circuit (Regulation 722.311)

A separate, dedicated electrical circuit shall be provided for the electric vehicle charging equipment or connecting point. However, more than one piece of electric vehicle charging equipment or connecting point may be fed from the same supply circuit, provided that the combined current demand of the equipment does not exceed the rating of the supply circuit.

6.10.3 Cable selection

The cable types and cross-sectional areas for the electric vehicle charging equipment circuit shall be chosen to ensure compliance with the requirements for:

(a) protection against electric shock (refer to Section 5.5);
(b) overcurrent protection (Chapter 43 of BS 7671:2018); and
(c) voltage drop (Section 525 of BS 7671:2018).

Where the final circuit supplies more than one charging point, no diversity shall be allowed.

Diversity may be allowed for a dedicated distribution circuit supplying multiple electric vehicle charging points if load control is available (Regulation 722.311).

When specifying the cable sizes for use in the electric vehicle charging equipment circuit, reference shall be made to either Appendix 4 of BS 7671:2018, which gives current-carrying capacity and voltage drop for cables, or manufacturer's data.

6.11 Protection against electric shock

As required by BS 7671, basic protection against electric shock shall be provided by automatic disconnection of supply or electrical separation.

6.11.1 Automatic disconnection of supply

Where automatic disconnection of the supply is used to provide protection against electric shock for the electric vehicle charging equipment circuit, compliance with Section 411 of BS 7671:2018 shall be ensured.

6.11.2 Additional protection

For additional protection, RCDs shall be fitted, as specified in Section 6.12, and supplementary equipotential bonding installed, as described in Section 6.9.

6.12 Requirements for the provision of RCDs (Regulation 722.531.2)

6.12.1 RCD specification

Every charging point shall be protected by a 30 mA RCD, with an operating time not exceeding 40 ms at a residual current of 5 $I_{\Delta n}$. Dedicated electric vehicle charging equipment should incorporate such an RCD, but the installer shall confirm this.

6.12.2 Selectivity

There should be selectivity between any RCD installed at the connecting point or incorporated in the charging equipment and the protection at the origin of the circuit (see Regulations 314.1 (i) and 536.4.1.4 of BS 7671:2018).

6.12.3 RCD protection (Regulation 722.531.2.101)

Every charging point shall be individually protected by an RCD having the characteristics specified in Regulation 415.1.1. The RCD shall disconnect all live conductors, which, in accordance with the definitions in Part 2 of BS 7671, includes the neutral conductor.

The RCD protecting the charging point shall be at least a Type A RCCB complying with BS EN 61008-1 or RCBO complying with BS EN 61009-1. If it is known that the DC component of the residual current exceeds 6 mA, then a Type B RCD that complies with BS EN 62423 shall be installed.

For charging in Mode 3 or 4, the equivalent of Type B RCD protection is required – either by use of a Type B RCD or by a Type A plus additional DC protection devices. Some charging equipment may incorporate DC fault protection and thus be compatible with a Type A RCD in the consumer unit.

For charging in Mode 2, Type A RCD protection is permitted, as DC fault protection is provided by the ICCB.

6.12.4 Labels

A label describing the operation and testing of the RCD, as shown in Figure 5.4, shall be fitted in a visible position, adjacent to the installed location of each device.

6.13 Requirements for isolation and switching

6.13.1 Emergency switch (Regulation 722.537.4)

Because all electric vehicle charging equipment and connecting points are required to be protected by a 30 mA RCD, an emergency switch is not usually required. However, the particular needs of the installation and the charging equipment manufacturer's instructions shall be considered. For instance, a readily accessible means of switching off the charging equipment shall be provided when the vehicle is to be charged within the confines of a domestic garage. If an emergency switch is provided, it shall be located in a position that is readily accessible, shall be suitably identified by marking and/or labelling and shall disconnect all live conductors including the neutral.

6.13.2 Isolation

A means of isolating the electric vehicle charging equipment circuit shall be provided, in accordance with Regulation 537.2 of BS 7671:2018. This isolating device shall be located in a position that is readily accessible for maintenance purposes and shall be suitably identified by marking and/or labelling.

6.13.2 Functional switching

Additional forms of switching for functional purposes may be required, for example to prevent the unauthorized use of socket-outlets located outside buildings that may be accessible to the public. These shall be discussed and agreed with the client.

6.14 IP ratings

6.14.1 IP rating of charging equipment (Regulation 722.512.2)

The electric vehicle charging equipment being installed shall have an IP rating suitable for the proposed installation location. For example, charging equipment that is to be installed in an outdoor location shall have an IP rating of at least IP44.

6.14.2 IP rating of installation components

The installer shall ensure that any connectors, glands, conduits, etc. used to connect the power supply to the electric vehicle charging equipment are of an appropriate IP rating, and are installed in such a way as to maintain a suitable IP rating, in accordance with BS 7671.

6.15 Lightning and surge protection systems

Where a building has a lightning protection system, reference shall be made to BS EN 62305 and Regulation 411.3.1.2 of BS 7671:2018.

Where surge protection is provided as part of the overall lightning protection system, cables serving charging points located outdoors may require additional surge protection.

Electrical requirements – On-street installations

This section contains specific requirements on the electrical installation of electric vehicle charging equipment at on-street locations.

7.1 ENA Engineering Recommendation G12

The requirements of the current Energy Networks Association (ENA) Engineering Recommendation G12 Issue 4 (incorporating Amendment 1) 2015 are as follows:

The IET would like to thank the ENA for permission to use this extract. This is copyright material and permission to further use this extract must be obtained from the ENA.

6.2.15 Street Electrical Fixtures not covered by 6.2.14

This Section covers roadside housings accessible to the public; examples are: cable television distribution cabinets, electric vehicle charging points, electrical distribution cabinets with load above 500W.

Note: A code of practice for electric vehicle charging equipment installation has been published by the IET.

Street electrical fixtures should preferably be of Class II construction or equivalent as defined in BS 7671. Examples are public telephones, pedestrian crossing bollards, ticket machines. No mains-derived earthing terminal is required, neither is a residual current device needed for earth fault protection.

Where the street electrical fixture is of Class I construction as defined in BS 7671, (examples include metal enclosures containing pumps, controls or communications equipment), a PME earth terminal may be provided if the requirements of BS 7671 are met and:

1 For 3-phase equipment, the load is balanced, or:
2 For 1-phase equipment and 3-phase equipment with unbalanced load, the maximum load and consumer earth electrode resistance bonded to their main earth terminal fulfil the requirements of Table 6.2.15.

Note: By agreement with the Network Operator, it may be permissible to take into account the effect of distributed earths in specific situations.

If the conditions for a Class I installation cannot be met, a PME terminal should not be offered. The earthing system of the installation should form part of a TT system by installing a separate earth electrode and fitting appropriate protection in accordance with BS 7671 (e.g. an RCD).

Extraneous-conductive-parts (e.g. safety barriers, pedestrian guard rails) should not be connected to a PME earth terminal.

The maximum consumer earth electrode resistances indicated in Table 6.2.15 of G12/4 allow slightly higher touch voltages than BS 7671. It is therefore recommended that the maximum earth electrode resistances given in Table 7.1 below are used.

▼ **Table 7.1 – Estimate of maximum consumer earth electrode resistance (R_A) for PME supplies (applying the equation in Annex A722 of BS 7671)**

Maximum 1-phase load or, for 3-phase, maximum overall load unbalance (kW)	I_m (A)	Maximum consumer earth electrode resistance allowed by BS 7671 (Ω)			
		Single-phase supplies	Three-phase supplies		
			$I_{L1}+I_{L2}+I_{L3}$ =10 A	$I_{L1}+I_{L2}+I_{L3}$ =30 A	$I_{L1}+I_{L2}+I_{L3}$ =100 A
0.5	2.1	46	NR	NR	NR
1	4.3	23	NR	NR	NR
2	8.7	11	NR	NR	NR
3	13.0	8	NR	NR	NR
4	17.4	6	23	NR	NR
5	21.7	5	11	NR	NR
7	30.4	3	6	NR	NR
14	60.9	1.6	2.1	5	NR
22	95.6	1	1.2	1.8	NR

NR earth electrode not required.

Note: For three-phase calculations it has been assumed the EV load in column 1 is added to the three-phase load $I_{L1}+I_{L2}+I_{L3}$.

7.2 General

Electric vehicle charging equipment should not be connected to an unmetered supply (as is common for street furniture). Due to the size of the load and its unpredictability, electric vehicle charging equipment should always be fed via a metered supply.

Tariff metering must be provided if the use of the charge point is billed on an energy usage basis, and may be required for energy management purposes, for example, where Part L2 of the Building Regulations apply.

For on-street installations, metering may be included in EVSE, or installed in a feeder pillar where there is insufficient space within the charging equipment.

7.3 Earthing and protective equipotential bonding requirements

On-street installations may be part of a TN-S or form part of a TT system. However, unless it can be guaranteed by the relevant DNO that the TN-S supply system is TN-S back to source and will not be converted to TN-C-S as part of any ongoing network upgrades, such installations shall be treated as TN-C-S systems (PME supply).

Distribution network operators will not provide the main earthing terminal of a PME supply for street furniture with loads greater than 2 kW, unless a consumer earth electrode is installed, that has a resistance to earth that complies with Table 6.2.15 of G12/4. BS 7671 requires lower consumer earth electrode resistances as per Table 7.1 above.

Consequently, when on-street electric vehicle charging equipment is to be connected to a TN-C-S (PME) supply, a separate TT earthing system must be provided for this charging equipment.

Before a TT system is installed, a risk assessment must be carried out so that risks of simultaneous contact with electrical installations, including other street furniture with PME earthing, are minimized (see Annex C). Generally, this will require the charging equipment to be installed in a location that ensures that this equipment and the vehicle on charge are at least 2.5 m from:

(a) other structures with exposed metalwork that are either connected to true Earth or connected to any other electrical earthing system; and

(b) any extraneous- or exposed-conductive-parts of any other item of electrical equipment.

Figure 7.1 shows a typical on-street electric vehicle charging equipment installation forming part of a TT system.

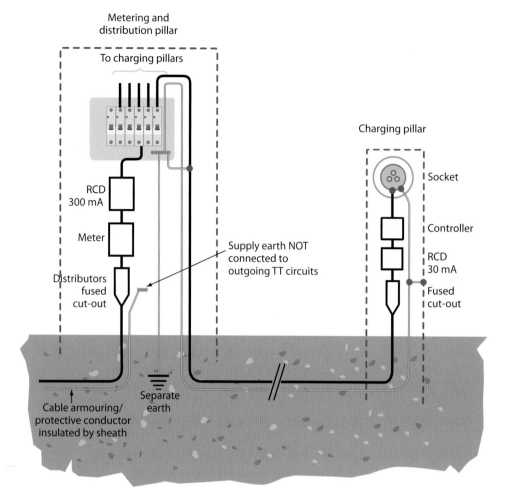

Figure 7.1 Typical on-street electric vehicle charging equipment installation forming part of a TT system

7.4 Requirements for circuit design, cable selection and dimensions

7.4.1 BS 7671 compliance

The electrical circuit for the electric vehicle charging equipment shall comply with the requirements of BS 7671, in particular, Section 722.

7.4.2 Circuit design

Whilst it is common for street lights and street furniture to be wired on a loop basis, this approach is not advisable for electric vehicle charging equipment. On-street charging equipment installations shall be wired as part of a radial system.

A separate, dedicated electrical circuit shall be provided for the electric vehicle charging equipment. However, more than one piece of electric vehicle charging equipment may be fed from the same supply circuit, provided that the combined current demand of the equipment does not exceed the rating of the supply circuit (Regulation 722.311).

7.4.3 Cable selection

The cable types and cross-sectional areas for the electric vehicle charging equipment circuit shall be chosen to ensure compliance with the requirements for:

(a) protection against electric shock (refer to Section 5.5);
(b) overcurrent protection (Chapter 43 of BS 7671:2018); and
(c) voltage drop (Section 525 of BS 7671:2018).

No diversity shall be allowed for final circuits, unless automatic controls limit the demand.

When specifying the cable sizes for use in the electric vehicle charging equipment circuit, reference shall be made to either Appendix 4 of BS 7671:2018, which gives current-carrying capacity and voltage drop for cables, or manufacturer's data.

7.4.4 Buried cables

WARNING

Excavating in the street is dangerous work: damage to underground services can cause fatal or severe injury. The activity of excavating in the street needs to be addressed in a safe manner. See the Health and Safety Executive (HSE) publication HSG47 *Avoiding dangers from underground services* (this publication is available as a free download at http://www.hse.gov.uk/pubns/priced/hsg47.pdf).

The work should be carried out by a specialist contractor.

Positioning and colour-coding of underground utilities' apparatus

The National Joint Utilities Group (NJUG) has agreed colours, depths of cover and positions for ducts, pipes, cables and marker/warning tapes when laid in the street. These are reproduced in Table 7.2 and Figure 7.2. This gives guidance to designers as to the depth at which cables might reasonably be expected to be laid generally, for example, in car parks, as well as in public paths and highways.

WARNING

The information provided in Table 7.2 and Figure 7.2 is current practice and indicates where it is preferable for newly installed cables and pipes to be installed. Older cables and pipes may not use these colour identifiers, nor be located at these depths. See HSE publication HSG47 for further guidance.

▼ **Table 7.2 – Recommended colour-coding of other underground apparatus**
Note: These guidelines are taken from the 'NJUG Publication Volume 1 –
NJUG Guidelines on the Positioning and Colour Coding of Underground Utilities'
Apparatus'

Asset Owner	Duct	Pipe	Cable	Marker Systems	Recommended Minimum Depths	
					Footway	Carriageway
Electricity HV (High Voltage)	Black or red duct or tile	N/A	Black or Red	Yellow with black and red legend or concrete tiles	450-1200 mm	750-1200 mm
Electricity LV (Low Voltage)					450 mm	600 mm
Gas	Yellow	*** See row below	N/A	Black legend on PE pipes every linear metre.	600 mm footway 750 mm verge	750 mm
	*** Pressure - up to 2 bar - yellow or yellow with brown stripes (removable skin revealing white or black core pipe). - between 2 to 7 bar -orange. Steel pipes may have yellow wrap or black tar coating or no coating. Ductile Iron may have plastic wrapping Asbestos & Pit / Spun Cast Iron - No distinguishable colour					
Water non Potable & Grey Water	N/A	Black with green stripes	N/A	N/A	600 - 750 mm	600 - 750 mm
Water Firefighting	N/A	Black with red stripes or bands	N/A	N/A	600 - 750 mm	600 - 750 mm
Oil / fuel pipelines	N/A	Black	N/A	Various surface markers Marker tape or tiles above red concrete	900 mm	900 mm
					All work within 3 m of oil / fuel pipelines must receive prior approval	*All work within 3 m of oil / fuel pipelines must receive prior approval*
Sewerage	Black	No distinguishing colour/material (e.g.: Ductile Iron may be red; PVC may be brown)	N/A	N/A	Variable	Variable
Telecomms	Grey, white, green, Black, purple	N/A	Black or light grey	Various	250 - 350 mm	450 - 600 mm
Water	Blue or Grey	Blue polymer or blue or uncoated Iron / GRP. Blue polymer with brown stripe *(removable skin revealing white or black pipe)*	N/A	Blue or Blue/black	750 mm	750 mm
Water pipes for special purposes (e.g. contaminated ground)	N/A	Blue polymer with brown stripe *(non-removable skin)*	N/A	Blue or Blue/black	750 mm	750 mm

Asset Owner	Duct	Pipe	Cable	Marker Systems	Recommended Min. Depths	
					Footway	Carriageway
Highway Authority Services						

Highway Authority Services
At the time of publication the following were current examples of known highway authority apparatus colour coding

Asset Owner	Duct	Pipe	Cable	Marker Systems	Footway	Carriageway
Street Lighting						
England and Wales	Black or Orange* * Consult electricity company first	N/A	Black	Yellow with black legend	450 mm	600 mm
Scotland	Purple	N/A	Purple	Yellow with black legend or purple	450 mm	
Northern Ireland	Orange	N/A	Black or Orange	Various	450 mm	450 mm
Other						
Traffic Control	Orange		Orange	Yellow with black legend		
Street Furniture	Black	N/A	Black	Yellow with black legend	450 mm	600 mm
Telecoms	Purple or Orange	N/A	Black	Various		
Motorways and Trunk Roads						
England and Wales						
Communications	Purple	N/A	Grey	Yellow with black legend	450 mm	
Communications Power	Purple	N/A	Black	Yellow with black legend		
Road Lighting	Orange	N/A	Black	Yellow with black legend		
Scotland						
Communications	Black or Grey	N/A	Black	Yellow with black legend		
Road Lighting	Purple	N/A	Purple	Yellow with black legend		

▼ Figure 7.2 National guidelines on the positioning of utility apparatus in a 2 m footway

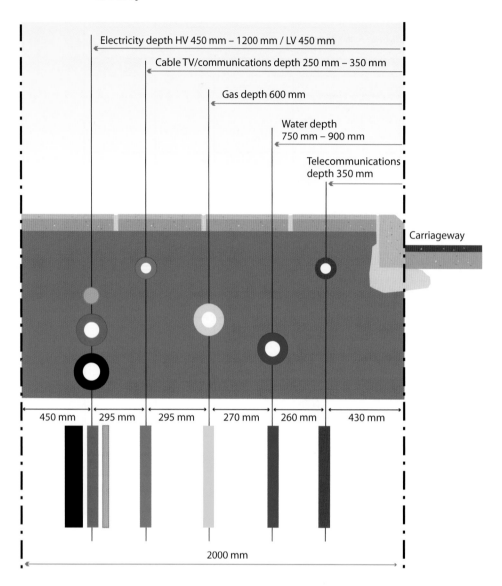

7.5 Requirements for the provision of RCDs (Regulation 722.531.2)

7.5.1 RCD specification

Every charging point shall be protected by a 30 mA RCD, with an operating time not exceeding 40 ms at a residual current of 5 $I_{\Delta n}$. Dedicated electric vehicle charging equipment should incorporate such an RCD, but the installer shall confirm this.

7.5.2 Selectivity

There shall be selectivity between any RCD installed at the connecting point or incorporated in the charging equipment and any other RCDs in the installation. For instance, where a 30 mA Type A RCD is installed in the charging equipment, a 100 mA, 300 mA or 500 mA Type S RCD could be fitted in the feeder pillar.

7.5.3 RCD protection (Regulation 722.531.2.101)

Every charging point shall be individually protected by an RCD having the characteristics specified in Regulation 415.1.1. The RCD shall disconnect all live conductors, which in accordance with the definitions of Part 2 of BS 7671, includes the neutral conductor.

The RCD protecting the charging point shall be at least a Type A RCCB complying with BS EN 61008-1 or RCBO complying with BS EN 61009-1. If it is known that the DC component of the residual current exceeds 6 mA then a Type B RCD that complies with BS EN 62423 shall be installed.

For charging in Modes 3 or 4, the equivalent of Type B RCD protection is required – either by use of a Type B RCD or by a Type A plus an additional DC protection device. Some charging equipment may incorporate DC fault protection and thus be compatible with a Type A RCD in the consumer unit.

For charging in Mode 2, Type A RCD protection is permitted, as DC fault protection is provided by the ICCB.

7.6 Requirements for isolation and switching

7.6.1 Emergency switch (Regulation 722.537.4)

Because all electric vehicle charging equipment and connecting points are required to be protected by a 30 mA RCD, an emergency switch is not usually required. However, the particular needs of the installation and the charging equipment manufacturer's instructions shall be considered. If an emergency switch is provided, it shall be located in a position that is readily accessible, shall be suitably identified by marking and/or labelling and shall disconnect all live conductors, including the neutral.

7.7 IP ratings

7.7.1 IP rating of charging equipment (Regulation 722.512.2.201)

The electric vehicle charging equipment being installed shall have an IP rating suitable for the installation location. For on-street installations, this means that charging equipment shall have an IP rating of at least IP44.

7.7.2 IP rating of installation components

The installer shall ensure that any connectors, glands, conduits, etc. used to connect the power supply to the electric vehicle charging equipment are of an appropriate IP rating, and are installed in such a way as to maintain a suitable IP rating, in accordance with BS 7671.

7.7.3 Impact

On-street electric vehicle charging equipment installations shall be protected against a minimum impact severity of AG2, as defined in Appendix 5 of BS 7671:2018.

SECTION 8

Electrical requirements – Commercial and industrial installations

This section contains specific requirements on the electrical installation of electric vehicle charging equipment at commercial and industrial locations. This includes installations in, or adjacent to, business premises, such as shops, offices, factories, and so on, and public and private car parks (single-level or multi-storey). These requirements are also applicable to installations in the car-parking areas of multiple-occupancy buildings, such as apartment blocks.

8.1 General

The installation arrangement that may be adopted in commercial and industrial installations is so wide that specific guidance cannot be provided. It is in commercial and industrial installations that private 11,000/400 V substations may be installed and where a TN-S supply may be available, simplifying the vehicle charging installation design and risk assessment.

Commercial and industrial installations, however small, must comply with sub-clauses (i), (ii) and (iii) of Regulation 722.411.4.1 (see Section 5.2.4).

8.2 Earthing and protective equipotential bonding requirements

8.2.1 TT systems

For electric vehicle charging equipment installations that form part of a TT system, the charging equipment may be connected to the existing earth whether the charging equipment is installed within a building or not. The installer must confirm that the earthing and bonding arrangements meet the requirements of BS 7671 for a TT system, and that any deficiencies are rectified.

© The Institution of Engineering and Technology

When there is an adjacent installation connected to a PME earth terminal, the vehicle charging equipment may be supplied from a TT supply, provided that:

(a) a risk assessment is undertaken to assess the possibility of simultaneous contact between any accessible conductive-parts connected to a PME earthing arrangement, and:
  (i) any conductive-parts connected to the main earthing terminal of the TT system, for example, the electric vehicle charging equipment, including the connecting point;
  (ii) the vehicle on charge.
(b) suitable precautionary measures are put in place to prevent all such identified means of simultaneous contact, for example, by inserting a length of plastic pipe into a metal pipe, or by replacing a Class I luminaire with a Class II luminaire, etc.

Notes:

(a) The written risk assessment document must in every case be included with the installation documentation (Installation Certificate, etc.). See Annex D.
(b) A suitable advisory label must be provided at the origin of every circuit supplied from a TT system stating that this must not be connected to a PME earthing facility.

8.2.2 TN-S systems

Unless it can be guaranteed by the relevant DNO that the TN-S supply system is TN-S back to source and will not be converted to TN-C-S as part of any ongoing network upgrades, such installations shall be treated as TN-C-S systems (PME supply).

For electric vehicle charging equipment installations that are part of a guaranteed TN-S system of an installation, for example, where the charging equipment is supplied via an 11 kV transformer at the installation, the charging equipment may be connected to the existing earthing arrangement whether the charging equipment is installed within a building or not. The installer must confirm that the earthing and bonding arrangements meet the requirements of BS 7671 for a TN-S system, and that any deficiencies are rectified.

8.2.3 PME supply

Inside buildings

For electric vehicle charging equipment installations supplied from installations with a PME supply (part of a TN-C-S system), the charging equipment may be connected to the PME earthing arrangements if the charging equipment is located inside the building and if it can be ensured that the vehicle being charged will always be within that same building (for example an underground car park, multi-storey car park, and so on). It is necessary to ensure that the protective bonding provisions comply with the requirements of BS 7671 where PME conditions apply. Note in particular Regulations 411.3.1.2 and 544.1.1.

It will also be necessary to take care to balance the loading on the three phases of the installation so as to reduce touch voltages in the event of an open circuit in the PEN conductor of the supply.

Outside buildings

If the electric vehicle charging equipment and/or the vehicle being charged are located outside the building, BS 7671 imposes the particular requirements of Regulation 722.411.4.1.

Three-phase supplies

Note: It is unlikely that Regulation 722.411.4.1(i) will be practicable in many real installations. It is strongly recommended that this approach is not used unless reliable load and load balance data is available (for instance, 6-12 month load profile accompanied by robust, reliable, calculations).

Regulation 722.411.4.1(i) can be met for three-phase supplies if the three phases are evenly loaded.

Regulation 722.411.4.1

(i) The charging point forms part of a three-phase installation that also supplies loads other than for electric vehicle charging and, because of the characteristics of the load of the installation, the maximum voltage between the main earthing terminal of the installation and Earth in the event of an open-circuit fault in the PEN conductor of the low voltage network supplying the installation does not exceed 70 V rms.

Note: Annex 722, item A722.2 gives some information relating to (i).

Annex A722 of BS 7671 provides calculations to be used where the loading on the phases is balanced; see also Table 7.1 in this Code of Practice.

If in doubt, the maximum voltage between the main earthing terminal of the installation and Earth in the event of an open-circuit fault in the PEN conductor of the low voltage network that supplies the installation must be calculated, to determine that it is unlikely to exceed 70 V rms.

▼ **Figure 8.1 TN-C-S system with open circuit PEN conductor**

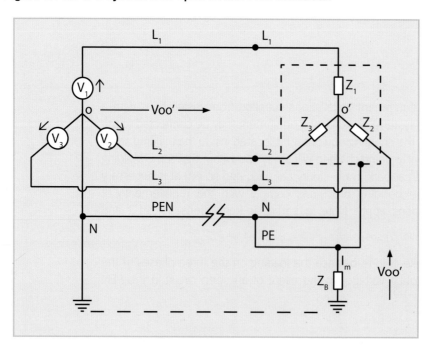

© The Institution of Engineering and Technology

▼ **Figure 8.2 Typical enclosed car park installation with multiple 13 A charging point connections**

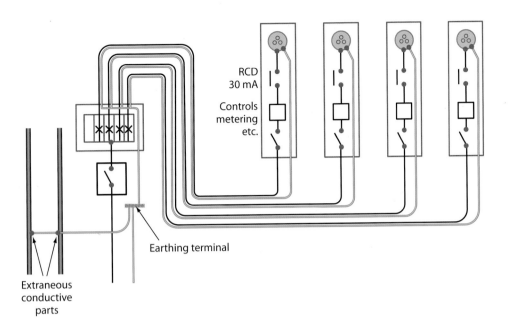

RCD
30 mA

Controls
metering
etc.

Earthing terminal

Extraneous
conductive
parts

Figure 8.3 shows a typical open-air car park installation utilizing a separate, dedicated TT earthing system.

▼ **Figure 8.3 Typical open-air car park installation utilizing a separate, dedicated TT earthing system**

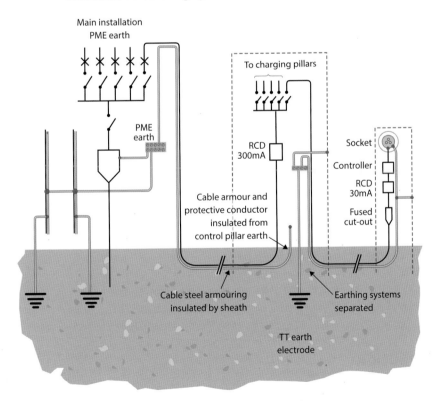

Main installation
PME earth

To charging pillars

PME
earth

RCD
300mA

Socket

Controller

RCD
30mA

Fused
cut-out

Cable armour and
protective conductor
insulated from
control pillar earth

Cable steel armouring
insulated by sheath

Earthing systems
separated

TT earth
electrode

8.3 Requirements for circuit design, cable selection and dimensions

8.3.1 BS 7671 compliance

The electrical circuit supplying the electric vehicle charging equipment shall comply with the requirements of BS 7671, in particular, Section 722.

8.3.2 Circuit design

Whilst it is common for street lights and street furniture is to be wired on a loop basis, this approach is not advisable for electric vehicle charging equipment. On-street charging equipment installations shall be wired as part of a radial system.

A separate, dedicated electrical circuit shall be provided for the electric vehicle charging equipment. However, more than one piece of electric vehicle charging equipment may be fed from the same supply circuit, provided that the combined current demand of the equipment does not exceed the rating of the supply circuit (Regulation 722.311).

8.3.3 Cable selection

The cable types and cross-sectional areas for the electric vehicle charging equipment circuit shall be chosen to ensure compliance with the requirements for:

(a) protection against electric shock (refer to Section 5.5);
(b) overcurrent protection (Chapter 43 of BS 7671:2018); and
(c) voltage drop (Section 525 of BS 7671:2018).

No diversity shall be allowed for final circuits, unless automatic controls limit the demand.

When specifying the cable sizes for use in the electric vehicle charging equipment circuit, reference shall be made to either Appendix 4 of BS 7671:2018), which gives current-carrying capacity and voltage drop for cables, or manufacturer's data.

8.4 Protection against electric shock

As required by BS 7671, basic protection against electric shock shall be provided by automatic disconnection of supply or electrical separation.

8.4.1 Automatic disconnection of supply

Where automatic disconnection of the supply is used to provide protection against electric shock for the electric vehicle charging equipment circuit, compliance with Section 411 of BS 7671:2018 shall be ensured.

8.4.2 Additional protection

Additional protection shall be provided by the fitment of RCDs as specified in Section 8.5.

8.5 Requirements for the provision of RCDs (Regulation 722.531.2)

8.5.1 RCD specification

Every charging point shall be protected by a 30 mA RCD, with an operating time not exceeding 40 ms at a residual current of 5 $I_{\Delta n}$. Dedicated electric vehicle charging equipment should incorporate such an RCD, but the installer shall confirm this.

8.5.2 Discrimination

There shall be discrimination between any RCD installed at the connecting point or incorporated in the charging equipment and the protection at the origin of the circuit.

8.5.3 RCD protection (Regulation 722.531.2.101)

Every charging point shall be individually protected by an RCD having the characteristics specified in Regulation 415.1.1. The RCD shall disconnect all live conductors, which, in accordance with the definitions of Part 2 of BS 7671, includes the neutral conductor.

The RCD protecting the charging point shall be at least a Type A RCCB complying with BS EN 61008-1 or RCBO complying with BS EN 61009-1. If it is known that the DC component of the residual current exceeds 6 mA then a Type B RCD that complies with BS EN 62423 shall be installed.

For charging in Modes 3 or 4, the equivalent of Type B RCD protection is required – either by use of a Type B RCD or by a Type A plus an additional DC protection device. Some charging equipment may incorporate DC fault protection and thus be compatible with a Type A RCD in the consumer unit.

For charging in Mode 2, Type A RCD protection is permitted, as DC fault protection is provided by the ICCB.

8.6 Requirements for isolation and switching

8.6.1 Emergency switch (Regulation 722.537.4)

Because all electric vehicle charging equipment and connecting points are required to be protected by a 30 mA RCD, an emergency switch is not usually required. However, the particular needs of the installation and the charging equipment manufacturer's instructions shall be considered. For instance, a readily accessible means of switching off the charging equipment shall be provided when the vehicle is to be charged within the confines of a single garage. If an emergency switch is provided, it shall be located in a position that is readily accessible, shall be suitably identified by marking and/or labelling and shall disconnect all live conductors, including the neutral.

8.6.2 Isolation

A means of isolating the electric vehicle charging equipment circuit shall be provided, in accordance with Regulation 537.2 of BS 7671:2018. This isolating device shall be located in a position that is readily accessible for maintenance purposes and shall be suitably identified by marking and/or labelling.

8.6.3 Functional switching

Additional forms of switching for functional purposes may be required, for example, to prevent the unauthorized use of socket-outlets located outside buildings that may be accessible to the public. These shall be discussed and agreed with the client.

8.7 IP ratings

8.7.1 IP rating of charging equipment (Regulation 722.512.2)

The electric vehicle charging equipment being installed shall have an IP rating suitable for the proposed installation location. For example, charging equipment that is to be installed in an outdoor location shall have an IP rating of at least IP44.

8.7.2 IP rating of installation components

The installer shall ensure that any connectors, glands, conduits, etc. used to connect the power supply to the electric vehicle charging equipment are of an appropriate IP rating, and are installed in such a way as to maintain a suitable IP rating, in accordance with BS 7671.

8.7.3 Impact

Electric vehicle charging equipment installed in publicly accessible locations, for example, car parks, shall be protected against a minimum impact severity of AG2, as defined in Appendix 5 of BS 7671:2018.

8.8 Lightning and surge protection systems

Where a building has a lightning protection system, reference shall be made to BS EN 62305 and Regulation 411.3.1.2 of BS 7671:2018.

Where surge protection is provided as part of the overall lightning protection system, any power and data cables serving charging points intended for charging vehicles outside existing premises may require surge protection devices.

Inspection, testing and maintenance requirements

9.1 BS 7671

During and on completion of the installation, and before being put into service, the installation shall be inspected and tested to verify that it complies with BS 7671 and the charging equipment manufacturer's instructions.

On completion, an Electrical Installation Certificate, together with schedules of inspection and schedules of test, shall be given to the client.

Periodic inspection and testing shall be carried out in accordance with BS 7671. Guidance on initial verification and periodic inspection and testing is given in IET Guidance Note 3: *Inspection & Testing*. Upon completion of a periodic inspection, an Electrical Installation Condition Report should be issued (see Regulation 653.1 and Appendix 6 of BS 7671:2018).

Inspection, testing and maintenance requirements for filling station installations should comply with the APEA/EI publication *Design, construction, modification, maintenance and decommissioning of filling stations*.

9.2 Electric vehicle charging equipment functional checks

9.2.1 General

The correct functioning of the charging equipment shall be verified.

Where controls from a filling station emergency switching system are provided, these shall be verified as operational.

9.3 Customer care

The correct operation of the charging equipment shall be demonstrated to the client.

9.4 Maintenance

The client shall be provided with the instruction manual for the equipment. Inform the client of any maintenance requirements for the equipment and of the need to add the equipment to any existing maintenance schedules.

9.5 Frequency of inspection and test

Charging equipment installations should be routinely checked, inspected and tested in accordance with the charging equipment manufacturer's recommendations. In the absence of any manufacturer's recommendations, inspection and testing should be undertaken with the frequency recommended in IET Guidance Note 3: *Inspection & Testing*, for the particular installation type.

For filling stations, refer to the APEA/EI publication *Design, construction, modification, maintenance and decommissioning of filling stations*.

9.6 RCD testing

The client shall be advised that all RCDs of the installation should be tested at least every six months (by pressing the test button).

SECTION 10

Vehicle as storage

10.1 Scope

This section concerns electric vehicles used for energy storage support, typically in single-family-occupancy dwelling installations.

Note: Installations where more than one vehicle is used to support energy storage would require additional, perhaps complex, design considerations, which are outside the scope of this Edition.

This section is intended to be read in conjunction with the IET *Code of Practice for Electrical Energy Storage Systems*.

10.2 Background

Electric vehicles often include considerably more energy storage capability than is required to meet day-to-day travel needs. This makes them a valuable resource as part of inherently large-scale distributed energy storage in the national energy system context.

With appropriate charge control equipment, electric vehicles may be charged or discharged at configurable power to:

(a) provide demand for low-cost surplus renewable generation. Charging is optimized to use either low-cost generation, or cheaper tariffs.
(b) reduce import from the grid supply during expensive peak periods, by using the stored energy in the vehicle to meet some of the installation demand when tariffs are expensive, and recharge the vehicle at a lower tariff.
(c) export electricity to the grid when the feed-in tariff is favourable.
(d) act as emergency back-up power during grid power cuts, when local renewable generation may be unavailable (for example, solar-photovoltaic systems having no output at night). Energy stored in the vehicle batteries can be used to power critical (typically low-power) loads for the duration of the power cut, and/or until local renewable generation sources begin to provide output.

10.3 Overall system design considerations

The following sections of the IET *Code of Practice for Electrical Energy Storage Systems* will help the designer to consider whether it is appropriate to provide vehicle as storage for a particular installation:

Section 3	System architectures
Section 4	System operating states and applications
Section 6	System components
Section 7	Safety and planning considerations
Section 8	Overall system specification, including sizing considerations
Section 9	Design for electrical safety

10.4 DNO notification for generation

DNO notification will be required when a vehicle as storage system that can operate in parallel with the grid supply is installed. The notification requirements are in accordance with ENA recommendations G83 or G59.

The actual requirements will depend on the total generation at the site that operates in parallel with the grid supply.

Section 10 of the IET *Code of Practice for Electrical Energy Storage Systems* contains decision trees and flow-charts that will assist in selecting the correct route for a particular installation.

10.5 Isolation and switching off for maintenance

Parts of the installation not capable of being isolated by a single device will require a warning notice in accordance with Regulation 537.1.2 of BS 7671. All isolators should have their function clearly marked in accordance with Regulation 537.2.7 of BS 7671.

10.5.1 Isolation and switching off for maintenance in dwellings

In a dwelling, there should ideally be a single point of isolation that removes power from all final circuits in the installation. For this type of installation supplied by single-phase supplies, BS 7671 requires a main isolator to disconnect all live conductors (i.e., both line and neutral).

10.5.2 Isolation and switching off for maintenance in other types of premises

The Electricity at Work Regulations 1989 apply. The HSE publication HSG85 *Electricity at work: Safe working practices* provides guidance on these matters.

As an electrical installation containing an electrical equipment safety system (EESS) will have parts of the installation supplied by at least two sources of energy, isolation procedures for those parts of the installation may not always be simple. It is imperative that those working on the installation can clearly and easily identify the device(s) required to isolate the part(s) of the installation they are working on.

10.6 Electrical considerations for parallel operation

10.6.1 What is parallel operation?

Parallel operation is where the inverter of the vehicle as storage charging equipment supplies power in parallel with the grid.

10.6.2 Design considerations

The IET *Code of Practice for Electrical Energy Storage Systems* recommends the designer to consider:

(a) integrating the storage generation into the installation so that it does not impact on the safety of the installation in parallel mode, and checking that RCDs are compatible with the inverters;

(b) the additional fault currents the inverter contributes, for both fault current protection and thermal effects (Chapters 42 and 43 of BS 7671); and

(c) the effects of the inverter on overload protection for circuits in the installation, in accordance with Regulation 551.7.2 of BS 7671.

10.7 Electrical considerations for island-mode operation

10.7.1 What is island-mode operation?

Island-mode operation is where the grid is disconnected, but at least some circuits are powered by local generation sources.

The designer will consider that the conditions and provisions for fault protection and overcurrent protection are likely to change, as the prospective fault current drops and earth fault loop impedances increase.

It should be noted that the grid supply may be unavailable because a fault within the installation itself has caused the operation of a supply-side protective device (for example, a service head fuse). Section 9.4.5 of the IET *Code of Practice for Electrical Energy Storage Systems* describes the designer's consideration. It recommends that island-mode operation should not be retro-fitted to installations where final circuits and distribution circuits protected by re-wireable fuses are powered during island-mode operation.

10.7.2 Back-up isolator

A back-up isolator arrangement is required that simultaneously performs the following functions:

(a) to isolate the grid supply, as the installation moves from parallel operation to island-mode operation; and

(b) to establish a neutral-earth bond (if required) just after, but at substantially the same time as, the grid supply is isolated.

When the back-up isolator switches, either to or from island-mode operation, parts of the installation may experience inrush currents.

10.7.3 Means of earthing

Generators in island mode cannot rely on the supplier's means of earthing, which may be disconnected for maintenance. The consumer therefore requires their own earth electrode, to comply with Regulation 551.4.3.2.1 of BS 7671.

In general, TN-S operation in island mode will help to minimize earth fault loop impedances. The switching operations required to establish TN-S in island mode, and the selection of appropriate consumer earth electrodes, should follow the IET *Code of Practice for Electrical Energy Storage Systems*.

Figure 10.1 provides an example of an arrangement for connecting vehicle as storage to a TN system. Figure 10.2 shows an example for a TT system.

10.7.4 Automatic disconnection

It is likely, particularly for dwellings and smaller commercial/industrial premises, that disconnection times can only be achieved by additional protection from 30 mA RCDs at the output of the inverter(s). It is important to check that any RCDs that are relied on for automatic disconnection in island mode can operate with DC components that can be present.

Where the grid supply is TN-C-S or TN-S, the earthing provisions for some circuits of special locations may well be separate from the rest of the installation. This could include the electric vehicle charging supply circuit (see Figure 10.1).

▼ **Figure 10.1 Example of connection of vehicle as storage to a TN system**

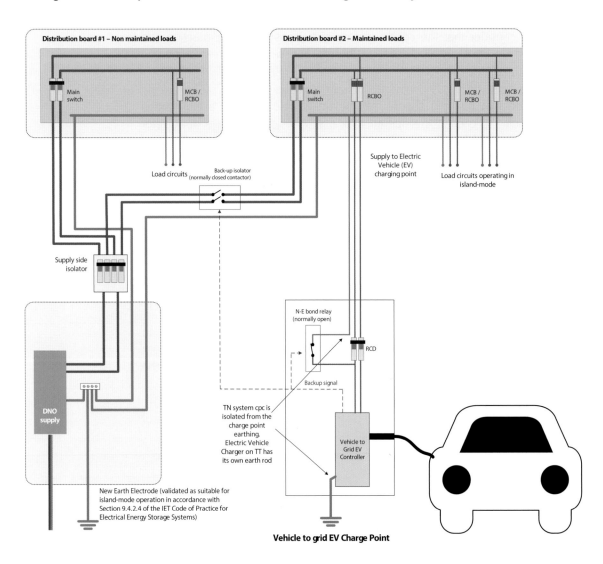

▼ **Figure 10.2 Example of connection of vehicle as storage to a TT system**

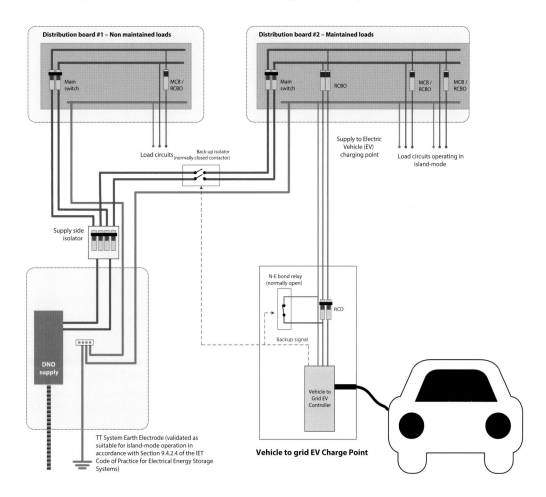

10.7.5 Protection against overcurrent and overload

The designer should consider whether back-up protection is required, as existing circuit protective devices may not operate due to reduced prospective fault currents.

10.7.6 Load-shedding

The local generation (vehicle as storage) inverters may not be able to support the demand for all loads in the installation. In addition, large loads will prematurely drain batteries.

Where the inverters and batteries are unlikely to support demand for the required period, load-shedding is recommended, so that only essential, low-power loads (for example, low-energy lighting, communications, etc.) are powered in island mode.

10.8 Multiple sources of generation

It is likely that the use of vehicle as storage in an installation forms part of an overall energy management and renewable generation strategy for a particular electrical installation. Where this is the case, the vehicle power converting equipment (PCE) may operate in parallel with other generators for renewable generation and/or energy storage.

Careful systems integration design is required to ensure that energy storage systems do not compete, reducing efficiency by increasing losses. For example, power from a vehicle as storage system should not normally be used to charge other energy storage systems in the installation.

All sources of generation that are connected (whether in parallel operation or island-mode operation) may contribute to faults.

Only a single neutral-earth bond should be in place at any one time. In parallel operation, DNO notification requirements are based on the total generation capability of the installation, not just the new generation added to the installation.

10.9   System labels and safety signs

Signage for multiple supplies will be required, in accordance with ENA Engineering Recommendations G83 or G59.

10.10 System handover and documentation

It is important that all users of the installation:

(a) understand whether the system is operating normally;

(b) know what alarms or error messages mean;

(c) are aware of the power limitations in island mode;

(d) understand the implications for the charge status of the vehicle and know how to prevent the vehicle from discharging where it may be needed for journeys; and

(e) know how to shut down the system to a safe state.

Additional information should be provided to ensure that the system can be safely maintained, inspected and tested.

SECTION 11

Distribution Network Operator (DNO) notification

In the case of dedicated electric vehicle charging equipment, the installer shall ensure that the appropriate DNO has been notified of the installation. Where the installation of charging equipment results in the property's **Maximum Demand (Load)** exceeding 13.8 kVA, an application to connect is required to be submitted to the relevant DNO prior to installation. This will enable DNOs to track increases in low-diversity demand from electric vehicle charging and thus mitigate the risks of distribution substation overloading and of voltages on the distribution network moving outside statutory limits.

The EV notification/application process and associated documents are subject to change. Installers should always access the ENA website for up-to-date information, instructions and the latest documents:
http://www.energynetworks.org/electricity/futures/electric-vehicle-infrastructure.html.

For Electric Vehicle Charging Equipment Installation, installers shall:

(a) where the **Maximum Demand (Load)**, including new Electric Vehicle Charging Equipment of a property is less than or equal to 13.8 kVA, the notification shall be sent by the installer within one calendar month of installation to EV-notifications@energynetworks.org, and ENA will forward to the appropriate DNO; or
(b) where the **Maximum Demand (Load)**, including new Electric Vehicle Charging Equipment of a property is greater than 13.8 kVA, or an issue has been identified with the adequacy or safety of the existing service equipment, the installer should contact the DNO prior to installation to establish the supply capacity.

Note that the notification and application form is designed to inform the DNO of installations of electric vehicle charging equipment. As described in Section 3, if the new load will exceed the existing supply, or if there is a planned programme of charging equipment installations in a close geographic region, the installer shall discuss the installation project with the local DNO at the earliest opportunity. The DNO will undertake an assessment of the impact that these connections may have on the network and then specify conditions for connection.

The Meter Point Administration Number (MPAN) or Meter Point Reference Number (MPRN) in Northern Ireland is essential in determining the location of the electric vehicle charging installation; this information can be found on the customer's bill or by contacting the energy supplier. Figure 11.1 details the numerical information that constitutes the MPAN.

▼ Figure 11.1 Example of MPAN information

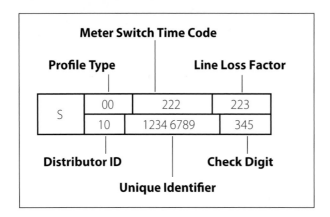

Integration and smart infrastructure

12.1 What is 'smart charging'?

The term 'smart charging' covers a number of features that can be addressed by various means of control. Examples (current and predicted) include:

(a) uni-directional (current through the charger flows to the car only):
  (i) timed charging control, to permit the vehicle to be charged at times when tariffs are typically lower;
  (ii) smart meter integration, permitting selection of charging options to ensure the car charges at the cheapest available rate, or enabling the charger to respond to dynamic tariffs;
  (iii) optimization for self-use (charging power is modulated to minimize export of generation to the grid and maximize self-consumption);
  (iv) demand-limiting charging, for example, high-power appliances in the same installation cause load shedding or throttling of the charger; and
  (v) managed charging, where electricity use by chargers is managed by the charge point operator (CPO) to defer DNO network upgrades.

(b) bi-directional (current may flow to or from the car):
  (i) vehicle to home (car as storage) used to avoid consumption during peak hours/peak rates;
  (ii) vehicle to grid: export from the car storage at peak demand, with car recharged more cheaply later; and
  (iii) off-grid systems: the vehicle may be used as storage in an overall off-grid system.

12.2 Means of control

Control by purely electromechanical means (relay or contactor switching) is not recommended. It introduces limitations in the system's flexibility and in the implementation of newer features as they become available over time. In some cases, the use of electromechanical control may cause charge control software to pause vehicle charging until the user intervenes (for example, acknowledging an alarm or status change).

Using a charger's micro-processor-based controller, in conjunction with web/smartphone application interfaces, smart meters and other controllers in the building, such as a building management system (BMS), supervisory control and data acquisition (SCADA) system or home automation systems (home and building electronic systems – HBES), provides far more flexibility and allows opportunities for future upgrades as more options become commercially available.

Table 12.1 provides some examples of control features and options that may be available to system designers. A system may exploit a number of features and means of control.

▼ Table 12.1 Examples of control features and options

Smart feature	Control type	Stand-alone control options	Integrated control options
Timed charging	Programmable/configurable	• Internal timed charging • Manual over-ride	• Mobile device applications or web-page (over the internet) for manual over-ride • External control from BMS, SCADA or HBES
Smart Meter	Programmable/configurable	• Manual over-ride	• SmartMeter interface to charge at cheapest rate • Mobile device applications or web-page (over the internet) for manual over-ride • Demand limiting based on overall current use in installation read from smart meter interface • Export optimization decisions
Demand limiting	Programmable/configurable	Manual over-ride	• Mobile device applications or web-page (over the internet) for manual over-ride • Mobile external control from BMS, SCADA or HBES • Smart meter interface permits vehicle charge controller to manage charge load based on installation demand
Optimization for self-use of local generation	Electromechanical	• Manual over-ride	• Smart meter interfaces are used to monitor generation and installation demand to govern vehicle charge control

12.3 Data integration planning considerations

12.3.1 Smart metering interfaces

Further information can be found in the BEAMA guide *Consumer Access Devices. Applications for Data in the Consumer Home Area Network (C HAN) and Wider Market Considerations.*

Information on becoming a registered Data Communications Company (DCC) user is available from the DCC website: www.smartdcc.co.uk.

Opportunities may exist in the future for integration with SMART Grid.

12.3.2 Protocol compatibility

When selecting interface protocols, it is necessary to consider the impact on:

(a) the cost of compatible devices to be integrated;

(b) the required competence and training for programming/configuration in commissioning and maintenance; and

(c) the requirements for physical interface cabling (or wireless interface infrastructure).

12.3.3 Privacy and information security

The Data Protection Act 1998, Regulation of Investigatory Powers Act 2000, and the Computer Misuse Act 1990 apply. The means of integration should consider both the privacy of individuals and information security.

It is strongly recommended that:

(a) any data gathered by commercial data collection systems that is provided to third parties for purposes other than billing and service provision is anonymized by design so that individuals and individual households cannot be identified;

(b) users and building owners are made aware of exactly what data is gathered and how that is going to be used, and that they provide their consent for that; and

(c) the means by which data is transmitted, accessed and encrypted is carefully considered, so that sensitive private or commercial information is not accessed without the correct authorization.

Further information can be found in the IET *Code of Practice for Connected Systems Integration in Buildings* and the BEAMA guide *Consumer Access Devices. Applications for Data in the Consumer Home Area Network (C HAN) and Wider Market Considerations.*

Information on becoming a registered DCC user for smart metering is available from the DCC website: www.smartdcc.co.uk.

12.4 Physical requirements for integration

12.4.1 Automation and control interfaces

Automation and control interfaces provide industry-standard communication protocols between devices that are optimized for control and monitoring purposes. Suitable interfaces may include:

(a) Metering Data, using one or more of the following:
  (i) smart metering:
  - BS EN 62056 series Electricity metering data exchange
  - BS EN 50491-11 HBES smart metering user display
  - PD CEN/CLC/ETSI TR 50572 Functional reference architecture for communications in smart metering systems

  (ii) PD CLC/TS 50568-4 and PD CLC/TS 50568-8 Electricity metering data exchange
  (iii) BS EN 61968-9 system interfaces for utility metering interface and control;

(b) Ethernet BS ISO/IEC/IEEE 8802-3-series standard for wired and wireless local area networks;

(c) ISO/IEC/IEEE 8802.15-4 standard for wireless personal area networks (e.g. Zigbee);

(d) KNX (BS ISO/IEC 14543-series compliant);

(e) LonWorks (BS EN 14908-series compliant);

(f) OPC (BS EN 62541-series);

(g) Profibus and Profinet (BS EN 62769-series);

(h) BACnet (BS EN 61484-5); and

(i) Modbus and Modbus over Ethernet.

12.4.2 Ethernet protocol for local area network and internet interfaces

Ethernet communication may be required for devices that connect to the home network, the internet (for example, to allow mobile device integration) or a corporate network for enterprise application integration.

Where the charger is required to communicate with devices on an Ethernet network, the first decision to make is whether to utilize a wired or wireless network.

Wired local area networks require suitable (CAT 5, 6 or 7, or fibre optic) physical connections to the local network or internet router or switch.

It should be considered whether the user knows how to maintain the network, should they need to replace any of their networking equipment.

Some of the considerations for network interfaces will be treated differently for installations using private data networks or broadband connections in dwellings, compared with those in infrastructure, commercial and industrial installations. In infrastructure, commercial and industrial installations, such differences may include, for example, the following points:

(a) electromagnetic compatibility (EMC) may be more of an issue with both wired and wireless networks.

(b) network management and security may be managed in line with corporate strategies and policies. This may drive the selection of particular products or protocols.

(c) the question as to whether there is sufficient wireless coverage for all devices to be connected may be considered.

(d) there may be more restrictions on shared networks.

(e) products may be required to operate using virtual private networks.

(f) devices may be required to be configurable to have fixed IP addresses, or use a specific domain name server.

(g) firewalls are likely to have greater restrictions.

(h) 'plug and play' type interfaces may have more restrictions or have firewall blocks preventing their use.

(i) remote device/system management and data logging may be prohibited from outside the corporate network.

12.4.3 General design considerations

12.4.3.1 Wired interfaces

Careful consideration should be given to electrical interference and EMC between power supply and power converting equipment, and the data interface cabling. The requirements of Section 444 of BS 7671 apply.

Fibre optic connections are less susceptible to EMC issues, but are more expensive.

Copper-wired networks are usually specified only to work within a building, according to BS EN 50174-2. Copper wiring to equipment outdoors, particularly at some distance from the building, should be considered in line with BS EN 50174-3. In addition, when equipment is copper-wired outdoors, the designer should consider whether the network equipment has the correct installation overvoltage category to prevent damage or fire from overvoltages. This is typically more of a problem for commercial and infrastructure installations and industrial sites. However, it may be an issue for installations associated with dwellings if the wiring extends some distance from the main building, and if there are utility, railway or other installations nearby that may provide overvoltage impulses.

12.4.3.2   Wireless interfaces

Wireless networks enable communication without physical wiring.

It is worth considering whether there is a suitable wireless transmission path for the devices to be connected. For example, are there any building materials between the devices that might block the transmission path? In addition, some phenomena may affect coverage at certain times or under certain conditions. Starting a boiler, for instance, or switching on fluorescent lights in the vicinity of the devices, might affect communications.

12.5 Software maintenance and lifecycle

Software (including firmware) often requires updating to address continuing functionality and to patch security flaws. This is particularly important for devices that may connect to the internet.

In dwellings, it is worth considering whether this is something the user/homeowner can or should be able to do themselves. For example, in systems with simple smartphone applications, the firmware/software in a controller may require updating for the applications to continue to operate.

In commercial and public charging installations, matching the firmware/software upgrade paths with the corporate IT policies and strategies is imperative. How change is notified by manufacturers/software providers is a key part of this strategy, as well as the time between notification and the requirement for the update to be applied. The selection of particular software vendors and equipment manufacturers may in fact be driven as much by the corporate IT requirements as by the functional requirements of the installation, and certain functional requirements may be curtailed accordingly.

12.6 Software failures and system recovery

Software control may not be able to perform normally for a number of reasons, for example:

(a) there has been a software error or crash;
(b) connectivity to a data provider or service becomes temporarily unavailable; or
(c) there is a hardware failure within communications or data-processing parts of the EVSE.

The system should be designed so that the ability to charge vehicles when this happens is not lost. Where software is used to manage loads, it may be necessary to limit the charge current (or number of chargers available) to prevent overload of the installation.

Performance in relevant failure modes should be considered very carefully to ensure electrical safety. For example, where controls are used to modulate the use of EVSE to prevent overloads of the installation (to prevent the system exceeding its maximum demand), the following might be necessary in a dwelling:

(a) when telemetry or communications that facilitate the function ceases, the software stops control of charging. The available charge current is curtailed completely or limited to prevent overload.
(b) to facilitate charging during failure, a manual switching arrangement is used as back-up, removing power from high-power loads (such as a shower, cooking appliances, etc.) equal to the maximum load of the charging system. The user can then opt to use either the EVSE or the other loads when automatic control is not available.

12.7 Electrical safety

Control systems should not lead to unsafe states. In particular, the following should be considered in the design of the installation and verified in commissioning:

(a) safe isolation. The control system should not be used for isolation for maintenance, inspection and testing. A manual means of isolation should be provided to facilitate safe working on all circuits, in accordance with BS 7671.
(b) protection against overload. Where the control system is used to modulate the use of EVSE to manage demand, relevant failure modes should be taken into account to prevent overload (which might in turn lead to equipment damage, fire, and so on). See Section 12.6 for examples.

Charging connectors and charging cable types

▼ Figure A.1 – Nomenclature of charging connection components

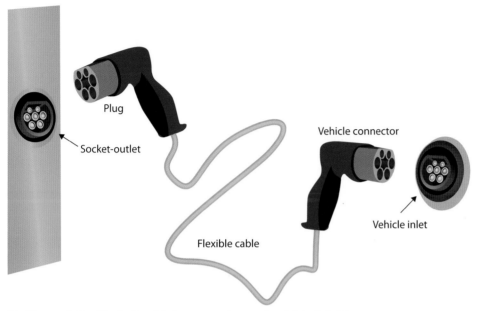

Plug

Socket-outlet

Vehicle connector

Vehicle inlet

Flexible cable

▼ Figure A.2 – Mode 3 vehicle connector and vehicle inlet types

BS EN 62196-2 Type 1 Vehicle Inlet and Vehicle Connector
Also known as the SAE J1772 connector or Yazaki connector

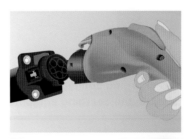

BS EN 62196-2 Type 2 Vehicle Inlet and Vehicle Connector
Also known as the Mennekes connector

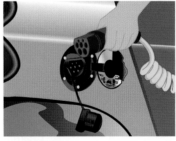

BS EN 62196-2 Type 3 Vehicle Inlet and Vehicle Connector
Also known as the EV Plug Alliance connector

▼ **Figure A.3.1 – Case A charging connection – charging cable and plug permanently attached to the electric vehicle**

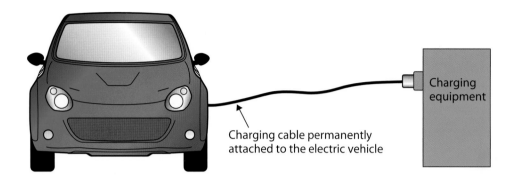

Charging cable permanently attached to the electric vehicle

Charging equipment

▼ **Figure A.3.2 – Case B charging connection – detachable charging cable assembly with a vehicle connector and plug**

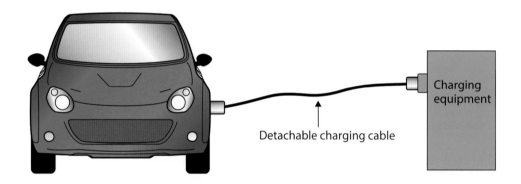

Detachable charging cable

Charging equipment

▼ **Figure A.3.3 – Case C charging connection – charging cable permanently attached to the charging equipment (tethered cable)**

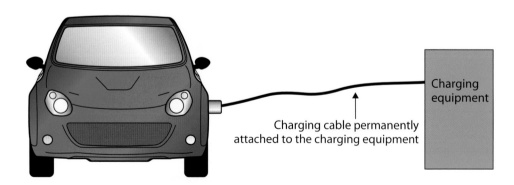

Charging cable permanently attached to the charging equipment

Charging equipment

▼ **Figure A.4 – Mode 4 Charging – charger with both combined charging system (CCS) and CHAdeMo tethered outlets**

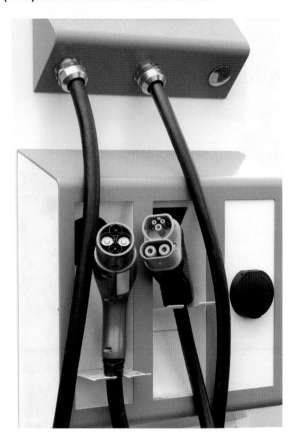

Checklists for dwelling installations

(Form to be included with forms for certification given to the person ordering the work.)

Arrangements prior to installation – Dwelling installations				
CoP Ref	**CHECK**	**YES**	**NO**	**N/A**
3.2	Is the existing supply adequate for the additional demand?			
3.3	Has the earthing arrangement of the incoming power supply been established?			
	Are the existing earthing and bonding arrangements compliant with BS 7671?			
6.5 and 6.6	Is the supply TN-C-S (PME) or TN-S?			
6.4	Is the supply TT?			
6.7 and 6.8	If TN-C-S or TN-S, have the precautions necessary been identified? (for example, isolating transformer, electrical separation, voltage operated trip, additional electrode system, adopting TT)?			
5.1.2 and 6.9	Has a simultaneous contact assessment been carried out? See form B1			
3.6	Has the installer reviewed the installation instructions provided by the charging equipment manufacturer?			
3.7	Has planning permission and/or Building Regulations approval been granted for the electric vehicle charging equipment installation?			
3.9	Have any constraints or difficulties of the proposed installation been discussed and agreed with the client?			
	Have any necessary repairs to the existing installation been agreed with the client?			

Physical installation requirements – Dwelling installations

CoP Ref	CHECK	YES	NO	N/A
4.2	Has the charging equipment been installed in an optimum location with respect to the intended vehicle parking position?			
4.3	Has the electric vehicle charging equipment been installed in a location to minimize the likelihood of vehicle impact damage?			
	If required, have protective barriers been provided?			
4.4	Are the main operating controls and any socket-outlets between 0.75 m and 1.2 m above the ground, and displays between 1.2 and 1.4 m above ground?			
4.5	Is there sufficient space around the charging equipment to open all doors and covers?			
4.6	Is there sufficient space around the charging equipment for ventilation and cooling purposes?			
4.7	Have all trip hazards been considered and, where possible, avoided?			
4.9	Have any BS 1363 socket-outlets been labelled as electric vehicle connecting points?			

Electrical installation requirements – Dwelling installations

CoP Ref	CHECK	YES	NO	N/A
	Pre-work survey of installation carried out including: • rating and condition of existing equipment • suitability for additional load • earthing and bonding			
	Pre-work tests of installation carried out including: • earth continuity, polarity and insulation resistance • earth fault loop impedance • operation of RCDs			
	Isolation of installation			
	Precautions taken to prevent inadvertent energizing			
	Defects in existing installation identified and notified to the client			
	Electrical Installation Certificate to hand, with preliminaries complete, including signatures for design			
	Installation isolated and precautions to prevent inadvertent switching on taken			
9	Preceding testing, inspection carried out on disconnected installation			
9	Inspections carried out as per BS 7671 Schedule of Inspections			
9	Schedule of Inspections completed			
9	Dead tests carried out as required by BS 7671 *prior to* energizing, and appropriate parts of the test schedule completed			
9	Remaining tests carried out as required by BS 7671 *after* energizing, and appropriate parts of the test schedule completed			
9	Electrical Installation Certificate completed, complete with schedule of inspections and schedule of test results			
9	Copy of certificates issued to the person ordering the work			
9	Customer advised in writing of any defects in the electrical installation not rectified			
9	Competent person scheme provider notified of completion			
9.3	Correct operation of the charging equipment demonstrated to the client			
9.4	Client provided with the instruction manual for the equipment and informed of any maintenance requirements			
11	DNO notification form for the installation submitted via the Energy Networks Association website			

Risk assessment form B1   IET Standards

Premise with PME supply and vehicle charging equipment to be installed outdoors where a TT system is proposed to be adopted for charging equipment only

(Form to be included with forms for certification given to the person ordering the work.)

Name and address of client	Installation Address
.... Postcode..	.... Postcode..
Supply taken from, e.g. main building/house/garage/other	....
Location of electric vehicle charging equipment considered in this risk assessment, (approximate distance to reference points)	....

Step	Record
Step 1 Identify the hazards	
(a) Does the building from which the charging supply is to be obtained have a PME (TN-C-S) or public TN-S supply?	Yes/No
(b) Is the vehicle charging equipment to be installed outdoors?	Yes/No
(c) Is a TT system to be adopted for the vehicle charging equipment only?	Yes/No *If the answer to all three of the above questions is* Yes, *the hazard will be:* In the event of an open-circuit neutral in the PME supply system, all conductive-parts connected to the main PME earthing terminal, e.g. any exposed- or extraneous-conductive-parts that may be directly, or indirectly, or otherwise connected to this earthing terminal, may become raised to a dangerous voltage relative to true Earth.
Step 2 Decide who might be harmed and how	Any person who can simultaneously touch any conductive-parts or conductor that might be connected to the main PME earth terminal of the building, e.g. a water tap, or metallic gas/water or fuel pipe, or metallic conduit, or item of Class I electrical equipment such as an outside light, switch or socket-outlet, or a boiler flue, or structural steel work, etc., AND the vehicle being charged OR any other conductive-parts or conductor that might be directly or indirectly or otherwise connected to the TT earth terminal of the vehicle charging equipment.

Step	Record
Step 3 Evaluate the risks and decide on precautions	
(1) Is it possible to simultaneously touch any conductive-parts or conductor that might be connected to the main PME (TN-C-S) or public TN-S earthing terminal AND the vehicle being charged OR any conductive-parts or conductor that might be connected to the TT earth terminal of the vehicle charging equipment? NB: All possible locations and positions of the vehicle on charge, and the charging lead and connector must be considered here.	Yes/No *If the answer to question 1 is No, retain the following text; if Yes, delete the following text:* 'THIS RISK ASSESSMENT SHOWS THAT IT IS NOT CURRENTLY NECESSARY TO TAKE ANY PRECAUTIONS TO PREVENT RISK OF SIMULTANEOUS CONTACT BETWEEN ANY CONDUCTIVE-PARTS OR CONDUCTOR THAT MIGHT BE CONNECTED TO THE MAIN PME EARTHING TERMINAL AND THE VEHICLE BEING CHARGED OR ANY OTHER CONDUCTIVE-PARTS OR CONDUCTOR THAT MIGHT BE CONNECTED TO THE TT EARTH TERMINAL OF THE VEHICLE CHARGING EQUIPMENT. THE CUSTOMER HAS BEEN INFORMED IN WRITING THAT A FURTHER RISK ASSESSMENT MUST BE UNDERTAKEN TO CHECK WHETHER THE CHARGING EQUIPMENT WILL REMAIN SAFE TO ENERGIZE AND/OR USE IF THIS SITUATION CHANGES.'
(2) If the answer to question (1) is Yes, can this simultaneous contact be reliably prevented, e.g. by fitting an insulating section into any pipe or conduit, or replacing any item of Class I equipment with Class II equipment, or by providing a permanent barrier or enclosure, or by applying permanent insulation, etc.?	Yes/No *If the answers to question 1 and question 2 are **both** Yes, record here the ESSENTIAL precautions required to prevent the possibility(ies) of simultaneous contact identified by Question (2):* .... *And* THE CUSTOMER HAS BEEN INFORMED IN WRITING THAT THE CHARGING EQUIPMENT WILL BECOME UNSAFE TO ENERGIZE AND/OR USE IF THE PRECAUTIONS LISTED ABOVE ARE REMOVED OR OTHERWISE BECOME INEFFECTIVE. *If the answer to question 1 is Yes and the answer to question 2 is No, retain the following text here (otherwise delete):* 'THIS RISK ASSESSMENT SHOWS THAT IT IS NOT CONSIDERED TO BE SAFE TO PROVIDE A TT EARTHED OUTDOOR VEHICLE CHARGING POINT AT THE CHOSEN LOCATION AND/OR TO CHARGE A TT EARTHED ELECTRIC VEHICLE AT THE CHOSEN LOCATION. THIS HAS BEEN MADE KNOWN TO THE CUSTOMER IN WRITING.'

Step	Record
Step 4 Record your findings and implement them If the precautions are inadequate, is it less of a risk to convert the complete installation to TT? If precautions are likely to be inadequate, consider the risks associated with the conversion of the complete installation to TT, considering, for example, adjoining properties on TN-C-S or TN-S, or adjoining properties with extraneous-conductive-parts or exposed-conductive-parts within reach of a vehicle on charge.	All precautions required by step 3 completed. Signature... Date carried out..
Step 5 Review the assessment and update if necessary	To be reviewed whenever further work is carried out on the installation, including any inspection and testing

Checklists for on-street installations

(Form to be included with forms for certification given to the person ordering the work.)

Arrangements prior to installation – On-street installations				
CoP Ref	**CHECK**	**YES**	**NO**	**N/A**
3.1	Has the installation of a new meter point been arranged with the DNO?			
7.3	Is the supply PME (TN-C-S) or a public TN-S supply?			
7.3	Is TT to be adopted for the installation?			
7.3	Has a simultaneous contact assessment been carried out? See risk assessment form C1.			
3.5	Has GPRS coverage of the proposed installation location been checked? *Note: Some EVSE manufacturers require a minimum of 3G coverage.*			
3.6	Has the installer reviewed the installation instructions provided by the electric vehicle charging equipment manufacturer?			
3.8	Has a Traffic Management Order (TMO) been created and duly consulted on for the electric vehicle charging equipment installation?			
3.9	Have the details of the proposed installation been discussed and agreed with the client?			

Physical installation requirements – On-street installations

CoP Ref	CHECK	YES	NO	N/A
4.2	Has the charging equipment been installed in an optimum location with respect to the intended vehicle parking position?			
4.3	Has the electric vehicle charging equipment been installed in a location to minimize the likelihood of vehicle impact damage?			
	If no, have protective barriers been provided?			
4.4	Has accessibility been addressed?			
	Are the main operating controls and any socket-outlets between 0.75 m and 1.2 m above the ground, and displays between 1.2 m and 1.4 m above the ground?			
	Have you made sure there are no changes of level between the vehicle and EVSE controls?			
	Is there adequate space for wheelchair access to the EVSE, controls and between two vehicles being charged?			
4.5	Is there sufficient space around the charging equipment to open all doors and covers?			
4.6	Is there sufficient space around the charging equipment for ventilation and cooling purposes?			
4.7	Have all trip hazards been considered and, where possible, avoided?			

Electrical installation requirements – On-street installations

CoP Ref	CHECK	YES	NO	N/A
	Pre-installation			
	Have New Roads and Street Works Act 1991 (NRSWA) notices been issued?			
7.4.4	Are drawings to hand of the electric cables and other services in installation locations?			
7.4.4	Have any potential hazards, e.g. high voltage (HV) cables, been identified in the excavation location?			
7.4.4	Has the electricity supply cable been identified, and the position of the distribution pillar been confirmed?			
	Has a permit to work been issued (where required), confirming that the installation is isolated?			
	Has the location of the earth electrode been identified and confirmed with the installer?			
	Has the location and depth of the trenches been confirmed?			
	Are arrangements in hand for permanent reinstatement of the cable routes?			
	Before connection			
9	Inspections carried out as per BS 7671 Schedule of Inspections			
	Schedule of Inspections completed			
	Dead tests carried out as required by BS 7671 *prior to* energizing, and appropriate parts of the test schedule completed			
	After supply connected			
	Remaining tests carried out as required by BS 7671 *after* energizing, and appropriate parts of the test schedule completed			
	Electrical Installation Certificate completed complete with schedule of inspections and schedule of test results			
Risk Assessment C1	Risk Assessment form C1 completed and given to client			
9.3	Correct operation of the charging equipment demonstrated to the client			
9.4	Client provided with the instruction manual for the equipment and informed of any maintenance requirements			
11	DNO notification form for the installation submitted via the Energy Networks Association website			

Risk assessment C1   IET Standards

On-street charging equipment, TT system

(Form to be included with forms for certification given to the person ordering the work.)

Name and address of client	Installation Address
....	....
....	....
....	....
Postcode..	Postcode..
Supply taken from, e.g. distribution cable/ service cable/building, specify	....
Location of electric vehicle charging equipment considered in this risk assessment, (approximate distance to reference points)	....

Step	Record
Step 1 Identify the hazards **Is the on-street charging equipment** installed in a location that will result in this equipment **and** the vehicle on charge being at least 2.5 m from: • other structures with exposed metalwork that are either connected to true Earth or connected to any other electrical earthing system? and • any extraneous- or exposed- conductive-parts of any other electrical equipment?	Yes/No *If the answer to the above questions is No, the hazard will be:* In the event of an open-circuit neutral in the PME supply system to street furniture or other installations, conductive-parts of such equipment may become raised to a dangerous voltage relative to the charging equipment or vehicle on charge.
Step 2 Decide who might be harmed and how	Any person who can simultaneously touch any conductive-parts of charging equipment or vehicle on charge and other street furniture or similar with an open-circuit neutral.

Step	Record
Step 3 Evaluate the risks and decide on precautions	
(1) Is it possible to simultaneously touch any conductive-parts of faulty street furniture AND the vehicle being charged OR any conductive-parts or conductor that might be connected to the TT earth terminal of the vehicle charging equipment? NB: All possible locations and positions of the vehicle on charge, and the charging lead and connector, must be considered here.	Yes/No *If the answer to question (1) is No, retain the following text; if Yes, delete the following text:* 'THIS RISK ASSESSMENT SHOWS THAT IT IS NOT CURRENTLY NECESSARY TO TAKE ANY PRECAUTIONS TO PREVENT RISK OF SIMULTANEOUS CONTACT BETWEEN ANY CONDUCTIVE-PARTS OR CONDUCTOR THAT MIGHT BE CONNECTED TO THE MAIN PME EARTHING TERMINAL AND THE VEHICLE BEING CHARGED OR ANY OTHER CONDUCTIVE-PARTS OR CONDUCTOR THAT MIGHT BE CONNECTED TO THE TT EARTH TERMINAL OF THE VEHICLE CHARGING EQUIPMENT. THE CUSTOMER HAS BEEN INFORMED IN WRITING THAT A FURTHER RISK ASSESSMENT MUST BE UNDERTAKEN TO CHECK WHETHER THE CHARGING EQUIPMENT WILL REMAIN SAFE TO ENERGIZE AND/OR USE IF THIS SITUATION CHANGES.'
(2) If the answer to question (1) is Yes, can this simultaneous contact be reliably prevented, e.g. by fitting an insulating section into any pipe or conduit, or replacing any item of Class I equipment with Class II equipment, or by providing a permanent barrier or enclosure, or by applying permanent insulation, etc.?	Yes/No *If the answers to question 1 and question 2 are both Yes, record here the essential precautions required to prevent the possibility(ies) of simultaneous contact identified by question (2):* .... *And* THE CUSTOMER HAS BEEN INFORMED IN WRITING THAT THE CHARGING EQUIPMENT WILL BECOME UNSAFE TO ENERGIZE AND/OR USE IF THE PRECAUTIONS LISTED ABOVE ARE REMOVED OR OTHERWISE BECOME INEFFECTIVE. *If the answer to question 1 is Yes and the answer to question 2 is No, retain the following text here (otherwise delete):* THIS RISK ASSESSMENT SHOWS THAT IT IS NOT CONSIDERED TO BE SAFE TO PROVIDE A TT EARTHED OUTDOOR VEHICLE CHARGING POINT AT THE CHOSEN LOCATION AND / OR TO CHARGE A TT EARTHED ELECTRIC VEHICLE AT THE CHOSEN LOCATION. THIS HAS BEEN MADE KNOWN TO THE CUSTOMER IN WRITING.

Step	Record
Step 4 Record your findings and implement them	All precautions required by step 3 completed Signature... Date carried out..
Step 5 Review the assessment and update if necessary	To be reviewed whenever further work is carried out on the installation including any inspection and testing.

Checklists for commercial and industrial installations

(Form to be included with forms for certification given to the person ordering the work.)

Arrangements prior to installation – Commercial and industrial installations				
CoP Ref	**CHECK**	**YES**	**NO**	**N/A**
4.1	Are there any hazardous zones where flammable/combustible gases may be present?			
4.1	Have the boundaries of any hazardous zones been identified?			
4.1	Can the installation be carried out so that the charged vehicle, cable and connectors are outside the hazardous area when charging?			
3.1	Is metering adequate for the intended use and billing model?			
3.2	Is the existing supply adequate for the additional demand?			
8	Has the earthing arrangement of the incoming power supply been established?			
	Are the existing earthing and bonding arrangements compliant with BS 7671?			
8.2.3	Is the supply PME (TN-C-S) or a TN-S public supply?			
8.2.3	If PME (TN-C-S), have precautions necessary to prevent danger in the event of an open-circuit neutral been identified and addressed?			
8.2.1	If a TT earthing system is being provided for the charging equipment, has a simultaneous contact assessment been carried out?			
3.5	Has GPRS coverage of the proposed installation location been checked? *Note: some EVSE manufacturers require a minimum of 3G coverage.*			
3.6	Has the installer reviewed the installation instructions provided by the charging equipment manufacturer?			
3.7	Has planning permission been granted for the charging equipment installation?			
3.9	Have the details of the proposed installation been discussed and agreed with the client?			
	Have any necessary repairs to the existing installation been agreed with the client?			

Physical installation requirements – Commercial and industrial installations

CoP Ref	CHECK	YES	NO	N/A
4.1	Has the charging equipment been installed outside any hazardous zones where flammable/combustible gases may be present?			
4.1	Have precautions been taken to ensure the charged vehicle, cable and connectors are outside the hazardous area when charging?			
4.2	Has the charging equipment been installed in an optimum location with respect to the intended vehicle parking position?			
4.3	Has the charging equipment been installed in a location to minimize the likelihood of vehicle impact damage?			
	If no, have protective barriers been provided?			
4.4	Are the main operating controls and any socket-outlets between 0.75 m and 1.2 m above the ground, and displays between 1.2 m and 1.4 m above ground?			
4.5	Is there sufficient space around the charging equipment to open all doors and covers?			
4.6	Is there sufficient space around the charging equipment for ventilation and cooling purposes?			
4.7	Have all trip hazards been considered and, where possible, avoided?			
4.9	Have any BS 1363 socket-outlets been labelled as electric vehicle connecting points?			

Electrical installation requirements – General

CoP Ref	CHECK	YES	NO	N/A
	Prior to installation checklist to hand and satisfactorily completed			
5, 8	Installation design to hand			
	Design section of electrical installation signed			
	Pre-work survey of installation carried out, including : • rating and condition of existing equipment • suitability for additional load • earthing and bonding			
	Pre-work tests of installation carried out, including: • earth continuity, polarity and insulation resistance • earth fault loop impedance • operation of RCDs			
	Defects in existing installation identified and notified to the client: • those affecting the new installation • those not affecting the new installation			
	Order to repair defects in existing installation affecting the new installation received			
	Equipment to be worked on isolated			
	Precautions taken to prevent inadvertent energizing			
	Installation carried out			
9	Testing			
9	Electrical Installation Certificate to hand, with preliminaries complete, including signatures for design			
9	Installation isolated and precautions to prevent inadvertent switching on taken			
9	Preceding testing, inspection carried out on disconnected installation			
9	Inspections carried out as per BS 7671 Schedule of Inspections			
9	Schedule of Inspections completed			
9	Dead tests carried out as required by BS 7671 *prior to* energizing, and appropriate parts of the test schedule completed			
9	Remaining tests carried out as required by BS 7671 *after* energizing, and appropriate parts of the test schedule completed			

Electrical installation requirements – General

CoP Ref	CHECK	YES	NO	N/A
9	Electrical Installation Certificate completed complete with schedule of inspections and schedule of test results			
9	Copy of certificates issued to the person ordering the work			
9	Customer advised in writing of any defects in the electrical installation not rectified			
9	Competent person scheme provider notified of completion			
9.3	Correct operation of the charging equipment demonstrated to the client			
9.4	Client provided with the instruction manual for the equipment and informed of any maintenance requirements			
11	DNO notification form for the installation submitted via the Energy Networks Association website			

Risk assessment D1 IET Standards

Installation with PME supply and vehicle charging equipment to be installed outdoors where a TT system is proposed to be adopted for charging equipment only

(Form to be included with forms for certification given to the person ordering the work.)

Name and address of client	Installation Address
.... Postcode..	.... Postcode..
Supply taken from, e.g. main intake, office unit z, store	....
Location of electric vehicle charging equipment considered in this risk assessment, (approximate distance to reference points)	....

Step	Record
Step 1 Identify the hazards	
(a) Does the building from which the charging supply is to be obtained have a PME supply?	Yes/No
(b) Is the vehicle charging equipment to be installed outdoors?	Yes/No
(c) Is a TT system to be adopted for the main installation and vehicle charging equipment?	Yes/No *If the answer to all three of the above questions is Yes, the hazard will be:* In the event of an open-circuit neutral in the PME supply system, all conductive-parts connected to the main PME earthing terminal, e.g. any exposed- or extraneous-conductive-parts that may be directly, or indirectly, or otherwise connected to this earthing terminal, may become raised to a dangerous voltage relative to true Earth.
Step 2 Decide who might be harmed and how	Any person who can simultaneously touch any conductive-parts or conductor that might be connected to the main PME earth terminal of the building, e.g. a water tap, or metallic gas/water or fuel pipe, or metallic conduit, or item of Class I electrical equipment, such as an outside light, switch or socket-outlet, or a boiler flue, or structural steel work, etc., AND the vehicle being charged OR any other conductive-parts or conductor that might be directly or indirectly or otherwise connected to the TT earth terminal of the vehicle charging equipment.

Step	Record
Step 3 Evaluate the risks and decide on precautions	
(1) Is it possible to simultaneously touch any conductive-parts or conductor that might be connected to the main PME earthing terminal AND the vehicle being charged OR any conductive-parts or conductor that might be connected to the TT earth terminal of the vehicle charging equipment? NB: All possible locations and positions of the vehicle on charge, and the charging lead and connector, must be considered here.	Yes/No *If the answer to question 1 is No, retain the following text; if Yes, delete the following text:* 'THIS RISK ASSESSMENT SHOWS THAT IT IS NOT CURRENTLY NECESSARY TO TAKE ANY PRECAUTIONS TO PREVENT RISK OF SIMULTANEOUS CONTACT BETWEEN ANY CONDUCTIVE-PARTS OR CONDUCTOR THAT MIGHT BE CONNECTED TO THE MAIN PME EARTHING TERMINAL AND THE VEHICLE BEING CHARGED OR ANY OTHER CONDUCTIVE-PARTS OR CONDUCTOR THAT MIGHT BE CONNECTED TO THE TT EARTH TERMINAL OF THE VEHICLE CHARGING EQUIPMENT. THE CUSTOMER HAS BEEN INFORMED IN WRITING THAT A FURTHER RISK ASSESSMENT MUST BE UNDERTAKEN TO CHECK WHETHER THE CHARGING EQUIPMENT WILL REMAIN SAFE TO ENERGIZE AND/OR USE IF THIS SITUATION CHANGES.'
(2) If the answer to question (1) is Yes, can this simultaneous contact be reliably prevented, e.g. by fitting an insulating section into any pipe or conduit, or replacing any item of Class I equipment with Class II equipment, or by providing a permanent barrier or enclosure, or by applying permanent insulation, etc.?	Yes/No *If the answers to question 1 and question 2 are both Yes, record here the essential precautions required to prevent the possibility(ies) of simultaneous contact identified by question (2):* .... *And* THE CUSTOMER HAS BEEN INFORMED IN WRITING THAT THE CHARGING EQUIPMENT WILL BECOME UNSAFE TO ENERGIZE AND/OR USE IF THE PRECAUTIONS LISTED ABOVE ARE REMOVED OR OTHERWISE BECOME INEFFECTIVE. *If the answer to question 1 is Yes and the answer to question 2 is No, retain the following text here (otherwise delete):* 'THIS RISK ASSESSMENT SHOWS THAT IT IS NOT CONSIDERED TO BE SAFE TO PROVIDE A TT EARTHED OUTDOOR VEHICLE CHARGING POINT AT THE CHOSEN LOCATION AND/OR TO CHARGE A TT EARTHED ELECTRIC VEHICLE AT THE CHOSEN LOCATION. THIS HAS BEEN MADE KNOWN TO THE CUSTOMER IN WRITING.'

Step	Record
Step 4 Record your findings and implement them If the precautions are inadequate, is it less of a risk to convert the complete installation to TT?	All precautions required by step 3 completed. Signature.. Date carried out..
Step 5 Review the assessment and update if necessary	To be reviewed whenever further work is carried out on the installation, including any inspection and testing.

Checklists for fuel filling station installations

(Form to be included with forms for certification given to the person ordering the work.)

Arrangements prior to installation – Fuel filling station installations				
CoP Ref	**CHECK**	**YES**	**NO**	**N/A**
4.1	Have the boundaries of any hazardous zones been identified?			
4.1	Can the installation be carried out so that the EVSE is located outside the hazardous area?			
4.1	Can the installation be carried out so that the charged vehicle, cable and connectors are outside the hazardous area when charging?			
3.1	Is metering adequate for the intended use and billing model?			
3.2	Is the existing supply adequate for the additional demand?			
8	Has the earthing arrangement of the incoming power supply been established?			
	Are the existing earthing and bonding arrangements compliant with BS 7671 and EI/APEA publication *'Guidance for Design, Construction, Modification, Maintenance and Decommissioning of Filling Stations'*?			
8.2.3	Is the supply PME (TN-C-S) or a TN-S public supply?			
8.2.3	If PME (TN-C-S), have precautions necessary to prevent danger in the event of an open-circuit neutral been identified and addressed?			
8.2.1	If required, has a simultaneous contact assessment been carried out?			
3.5	Has GPRS coverage of the proposed installation location been checked? *Note: Some EVSE manufacturers require a minimum of 3G coverage. GPRS must not affect safety of hazardous areas.*			
3.6	Has the installer reviewed the installation instructions provided by the charging equipment manufacturer?			
3.7	Has planning permission been granted for the charging equipment installation?			
3.9	Have the details of the proposed installation, and associated risk assessments, been discussed and agreed with the client?			
	Have any necessary repairs to the existing installation been agreed with the client?			

Physical installation requirements – Fuel filling station installations

CoP Ref	CHECK	YES	NO	N/A
4.1	Has the charging equipment been installed outside any hazardous zones where flammable/combustible gases may be present?			
4.1	Have precautions been taken to ensure the charged vehicle, cable and connectors are outside the hazardous area when charging?			
4.1.2	Where a separate utility company supply is provided for the EVSE, has a prominent warning label been mounted on the supply cubicle to indicate that the charger is fed from this separate supply and is not controlled by the filling station main switch?			
4.1.2	Has control for the EVSE supply by the forecourt emergency switching system been provided?			
4.1.2	If control by the emergency switching system has not been provided, because a separate utility supply has been used for the EVSE, has a prominent label been provided on the EVSE to indicate that it is not controlled by the filling station main switch? *Note: The EI/APEA publication 'Guidance for Design, Construction, Modification, Maintenance and Decommissioning of Filling Stations' requires the filling station emergency switching to control the supply to the EVSE.*			
4.2	Has the charging equipment been installed in an optimumlocation with respect to the intended vehicle parking position?			
4.1.2	Does the EVSE have tethered cable only?			
4.3	Has the charging equipment been installed in a locationto minimize the likelihood of vehicle impact damage?			
	If no, have protective barriers been provided?			
4.4	Are the main operating controls and any socket-outlets between 0.75 m and 1.2 m above the ground, with displays 1.2 m to 1.4 m above the ground?			
4.5	Is there sufficient space around the charging equipment to open all doors and covers?			
4.6	Is there sufficient space around the charging equipment for ventilation and cooling purposes?			
4.7	Have all trip hazards been considered and, where possible, avoided?			

Electrical installation requirements – General

CoP Ref	CHECK	YES	NO	N/A
	Prior to installation checklist to hand and satisfactorily completed			
5, 8	Installation design to hand			
	Design section of electrical installation signed			
	Pre-work survey of installation carried out, including: • rating and condition of existing equipment • suitability for additional load • earthing and bonding			
	Pre-work tests of installation carried out, including: • earth continuity, polarity and insulation resistance • operation of RCDs *Note: Earth fault loop impedance testing is not permitted by the EI/APEA publication 'Guidance for Design, Construction, Modification, Maintenance and Decommissioning of Filling Stations' and therefore it will be necessary to determine the earth fault loop impedance by other means as detailed in the EI/APEA publication and to conform with BS 7671..*			
	Defects in existing installation identified and notified to the client: • those affecting the new installation • those not affecting the new installation			
	Order to repair defects in existing installation affecting the new installation received, or, where necessary, provided to a contractor competent to work in hazardous areas of filling stations.			
	Equipment to be worked on isolated			
	Precautions taken to prevent inadvertent energizing			
	Installation carried out			
9	Testing			
9	Electrical Installation Certificate to hand, with preliminaries complete, including signatures for design			
9	Installation isolated and precautions to prevent inadvertent switching on taken			
9	Preceding testing, inspection carried out on disconnected installation			
9	Inspections carried out as per schedule of inspections and testing outlined in the EI/APEA publication *'Guidance for Design, Construction, Modification, Maintenance and Decommissioning of Filling Stations'*			

Electrical installation requirements – General

CoP Ref	CHECK	YES	NO	N/A
9	Schedule of Inspections completed			
9	Dead tests carried out as required by schedule of inspections and testing outlined in the EI/APEA publication 'Guidance for Design, Construction, Modification, Maintenance and Decommissioning of Filling Stations'			
9	Remaining tests carried out as required by the EI/APEA publication 'Guidance for Design, Construction, Modification, Maintenance and Decommissioning of Filling Stations'			
9	Electrical Installation Certificate completed complete with schedule of inspections and schedule of test results			
9	Copy of certificates issued to the person ordering the work			
9	Customer advised in writing of any defects in the electrical installation not rectified			
9	Competent person scheme provider notified of completion			
9.3	Correct operation of the charging equipment demonstrated to the client			
9.4	Client provided with the instruction manual for the equipment and informed of any maintenance requirements			
11	DNO notification form for the installation submitted via the Energy Networks Association website			

Risk assessment E1 IET Standards

Filling station electrical installations

(Form to be included with forms for certification given to the person ordering the work.)

Name and address of client	Installation Address
.... Postcode..	.... Postcode..
Supply taken from, e.g. main intake, separate DNO supply	....
Location of electric vehicle charging equipment considered in this risk assessment (approximate distance to reference points)	....

Step	Record
Step 1 Identify the hazards (a) Is any part of the filling station from which the charging supply is to be obtained earthed separately to the hazardous area TT supply?	Yes/No
(b) Is a TT system to be adopted for the main installation and vehicle charging equipment? Note: EVSE will almost always be TT at a filling station. Occasionally, site supplies, including the hazardous area, may be found to be PME, or TN-S from a dedicated transformer. Either the entire site will be provided with a TT earthing system, or none of the site will. The hazardous area should never be a different earthing system to the rest of the site.	Yes/No *If the answer to both of the above questions is Yes, the hazard will be:* First: in the event of an open-circuit neutral in the PME supply system, all conductive-parts connected to the main PME earthing terminal, e.g. any exposed- or extraneous-conductive-parts that may be directly, or indirectly, or otherwise connected to this earthing terminal, may become raised to a dangerous voltage relative to true Earth. Second: in general, equipment in the hazardous area of the filling station should NOT be supplied by TN-C-S (PME) supplies: THIS SHOULD NOT HAPPEN AT A FILLING STATION. The implications of this MUST be addressed with the person responsible for the safety of the filling station.
Step 2 Decide who might be harmed and how	Any person who can simultaneously touch any conductive-parts or conductor that might be connected to the main PME earth terminal of the building, e.g. a water tap, or metallic gas/water or fuel pipe, or metallic conduit, or item of Class I electrical equipment, such as an outside light, switch or socket-outlet, or a boiler flue, or structural steel work, etc., AND the vehicle being charged OR any other conductive-parts or conductor that might be directly or indirectly or otherwise connected to the TT earth terminal of the vehicle charging equipment. Anyone affected by explosion or ignition risks in the hazardous area, should the vehicle on charge be located in the hazardous area, or the charging equipment located in the hazardous area.

Step	Record
Step 3 Evaluate the risks and decide on precautions	
(1) Is it possible to simultaneously touch any conductive-parts or conductor that might be connected to the main PME earthing terminal AND the vehicle being charged OR any conductive-parts or conductor that might be connected to the TT earth terminal of the vehicle charging equipment? NB: All possible locations and positions of the vehicleon charge, and the charging lead and connector, must be considered here.	Yes/No *If the answer to question 1 is No, retain the following text; if Yes, delete the following text:* 'THIS RISK ASSESSMENT SHOWS THAT IT IS NOT CURRENTLY NECESSARY TO TAKE ANY PRECAUTIONS TO PREVENT RISK OF SIMULTANEOUS CONTACT BETWEEN ANY CONDUCTIVE-PARTS OR CONDUCTOR THAT MIGHT BE CONNECTED TO THE MAIN PME EARTHING TERMINAL AND THE VEHICLE BEING CHARGED OR ANY OTHER CONDUCTIVE-PARTS OR CONDUCTOR THAT MIGHT BE CONNECTED TO THE TT EARTH TERMINAL OF THE VEHICLE CHARGING EQUIPMENT. THE CUSTOMER HAS BEEN INFORMED IN WRITING THAT A FURTHER RISK ASSESSMENT MUST BE UNDERTAKEN TO CHECK WHETHER THE CHARGING EQUIPMENT WILL REMAIN SAFE TO ENERGIZE AND/OR USE IF THIS SITUATION CHANGES.'
(2) If the answer to question (1) is Yes, can this simultaneous contact be reliably prevented, e.g. by fitting an insulating section into any pipe or conduit, or replacing any item of Class I equipment with Class II equipment, or by providing a permanent barrier or enclosure, or by applying permanent insulation, etc.?	Yes/No *If the answers to question 1 and question 2 are both Yes, record here the essential precautions required to prevent the possibility(ies) of simultaneous contact identified by question (2):* .... *And* THE CUSTOMER HAS BEEN INFORMED IN WRITING THAT THE CHARGING EQUIPMENT WILL BECOME UNSAFE TO ENERGIZE AND/OR USE IF THE PRECAUTIONS LISTED ABOVE ARE REMOVED OR OTHERWISE BECOME INEFFECTIVE. *If the answer to question 1 is Yes and the answer to question 2 is No, retain the following text here (otherwise delete):* 'THIS RISK ASSESSMENT SHOWS THAT IT IS NOT CONSIDERED TO BE SAFE TO PROVIDE A TT EARTHED OUTDOOR VEHICLE CHARGING POINT AT THE CHOSEN LOCATION AND/OR TO CHARGE A TT EARTHED ELECTRIC VEHICLE AT THE CHOSEN LOCATION. THIS HAS BEEN MADE KNOWN TO THE CUSTOMER IN WRITING.'

Step	Record
Step 4 Record your findings and implement them	All precautions required by step 3 completed. Signature... Date carried out..
Step 5 Review the assessment and update if necessary	To be reviewed whenever further work is carried out on the installation including any inspection and testing.

ANNEX F

Wireless power transfer (WPT) installations

Overview

WPT charging is a technology currently in development. It will offer the ability to charge vehicle batteries without physical wiring between the vehicle and the charging equipment. Charging will instead be carried out wirelessly, for example, via magnetic field (inductive charging) or microwave.

At the time of publication of the current edition of this Code of Practice, manufacturers are advertising that WPT charging solutions are close to production. It is therefore likely that we will see a requirement to install related charging equipment within the lifecycle of this edition of the Code of Practice.

This Annex provides some guidance on how this Code of Practice might be used to support inductive charging installations.

Standards

There are no currently available standards that specifically relate to the installation of WPT equipment. The BS EN 61980 series of standards is being developed to support the technology.

Considerations for WPT charging equipment

In general, this Code of Practice applies equally to WPT and conductive charging installations. It is anticipated that the control and power requirements will be similar to Mode 3 charging systems, but of course, plug and socket connections will not be used. Some proposed systems use a paddle and cable to facilitate the power transfer.

If the WPT system has neither socket-outlets nor exposed-conductive-parts, then supply from a PME system may be appropriate, particularly in cases where the installation is associated with single dwelling.

Particular risks and issues relating to WPT installations are discussed in Table F1.

▼ **Table F1 – Particular risks and issues relating to WPT installations**

Risk/Issue	Recommendations	Standards and guidance
Underground services and buried cables	Cabling to inductive charging equipment shall be adequately routed and protected. Excavation work and buried services coordination may be necessary; Section 7.4.4 of this Code of Practice should be followed.	BS 7671:2018
Electromagnetic compatibility	The manufacturer's instructions relating to proximity of inductive charging equipment to other electrical equipment, and wiring of the electrical installation, shall be followed.	BS 7671:2018 (in particular, Section 444) Electromagnetic Compatibility Regulations 2016 BS EN 61000-6 series
Electromagnetic fields	The effects of electromagnetic fields on human beings and livestock shall be considered. The installer shall use manufacturer's information to ascertain the risks, and follow the manufacturer's recommendation for any required monitoring, controls and interlocks. In places of work, the installer shall liaise with the relevant duty-holder in relation to their obligations under the Control Of Electromagnetic Fields at Work Regulations 2016	HSE Guidance HSG281 *A guide to the Control of Electromagnetic Fields at Work Regulations 2016* Demonstration of compliance of equipment and exposure in the workplace: • BS EN 50664 • BS EN 50499 Assessment of electrical and electronic equipment related to human exposure restrictions: • BS EN 50665 • BS EN 62311 Measurement and calculation procedures for human exposure: • BS EN 50413 • BS EN 50500

Installing an earth electrode system to enable use of a PME supply earth

Estimation of resistance to earth

The resistance of an earth electrode system can be estimated as follows. Resistance (R_r) of one vertical electrode is given by:

$$R_r = \frac{\rho}{2\pi L}\left[\log_e\left(\frac{8L}{d}\right) - 1\right]$$

where:

R_r is the resistance of single rod or pipe, in ohms (Ω);
L is the length of rod, in metres (m);
d is the diameter of rod or pipe, in metres (m);
ρ is the soil resistivity in ohm metres (Ωm).

The resistance of n parallel-aligned vertical electrodes can be estimated using:

$$R_t = \frac{1}{n}\frac{\rho}{2\pi L}\left[\log_e\left(\frac{8L}{d}\right) - 1 + \frac{\lambda L}{s}\right]$$

where:

$$\lambda = 2\sum\left(\frac{1}{2} + \ldots + \frac{1}{n}\right),$$

s is the rod spacing, in metres (m).

For further arrangements, see BS 7430 *Code of practice for protective earthing of electrical installations*.

Tables G1 and G2 provide estimates of electrode system resistance (R_t) for assumed soil resistivities (ρ) and number of earth rods (n).

▼ **Table G1 Estimate of earth electrode system resistance (Rt) for an earth rod of length (L) 2.4 m and rod spacing(s) of 3 m**

Number of rods (n)	Soil resistivity (ρ)				
	10	20	50	100	1000
1	4	8	20	40	400
2	2.3	4.6	11	23	228
3	1.6	3.3	8.2	16	163
4	1.3	2.6	6.4	13	129
5	1.1	2.1	5.3	11	107

▼ Table G2 Estimate of earth electrode system resistance (Rt) for an earth rod of length (L) 1.2 m and rod spacing(s) of 3 m

Number of rods (n)	Soil resistivity (ρ)				
	10	20	50	100	1000
1	7.1	14	35	71	715
2	3.3	7.6	19	38	384
3	2.7	5.4	13	27	267
4	2.1	4.2	10	21	207
5	1.7	3.4	8	17	170

Required resistance to earth

The earthing system resistance (rod electrode, bare cable, etc.) required for a given installation current is given by:

$$R_{A\,ev} = \frac{70 C_{max} U_0}{I_{inst}(C_{max} U_0 - 70)}$$

$$= \frac{17710}{I_{inst} \times 183}$$

where:

I_{inst} is the rms maximum demand current of a single-phase installation (in amps), including that of the electric vehicle charging load and any other loads, determined in accordance with Regulation 311.1;

$R_{A\,ev}$ is the sum of the resistances of the earth electrode and the protective conductor connecting it to the main earthing terminal of the installation (in ohms), in item A722.3;

U_o is the nominal AC rms line voltage to Earth; and

C_{max} is the voltage factor of 1.1 for a supply derived in accordance with the ESQCR.

The maximum values of $R_{A\,ev}$ for given I_{inst} are shown in Table G3.

The installation maximum demand current is the sum of the maximum electric vehicle charging current plus the simultaneous demand of the rest of the installation. Although the vehicle maximum charging current will be known, the demand of the remainder of the installation is difficult to estimate.

Table G3 assumes an installation demand of 2 kW.

▼ Table G3 Maximum electrode resistance ($R_{A\,ev}$) required to limit the touch voltage to 70 V for single-phase supplies

Electric vehicle charging current (amps)*	$R_{A\,ev}$ (ohms)
10	5.1
20	3.3
32	2.4
60	1.4

* Assumes an installation current of 2 kW (9 A) excluding the vehicle charging current.

Tables G1, G2 and G3 together demonstrate that in many installations it will not be practicable to achieve the required electrode system resistances to earth.

However, it is recommended that a minimum of two 2.4 m or three 1.2 m electrodes separated by 3 m are installed.

Figure G.1 shows an example schematic using two 2.4 m length electrodes separated by 3 m.

▼ **Figure G.1 Typical domestic PME supply schematic with additional earth electrodes**

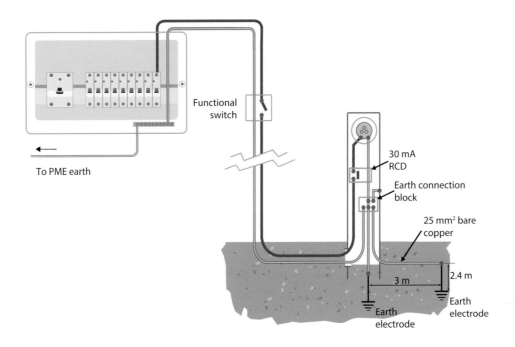

Note: Neutral omitted for clarity

Glossary

Client	In the context of this Code of Practice, 'client' means the designated representative of the person, company or body that has requested the electric vehicle charging equipment installation.
Competent person	In the context of this Code of Practice, 'competent person' means a person who possesses sufficient technical knowledge, relevant practical skills and experience for the nature of the electrical work undertaken and who is able, at all times, to prevent danger and, where appropriate, injury to him/herself and others.
Electric vehicle charging point	The point where the electric vehicle is connected to the fixed installation. **Note:** The charging point is a socket-outlet where the charging cable belongs to the vehicle, or a connector where the charging cable is a fixed part of the electric vehicle supply equipment.
Exposed-conductive-part	Conductive part of equipment that can be touched and that is not normally live, but can become live under fault conditions. For further guidance, refer to BS 7671 and IET Guidance Notes 1, 3, 5 and 8.
Extraneous-conductive-part	Conductive part liable to introduce a potential, generally Earth potential, and not forming part of the electrical installation. For further guidance, refer to BS 7671 and IET Guidance Notes 1, 3, 5 and 8.
DNO	Distribution network operator
Electric Vehicle	In the context of this Code of Practice, 'electric vehicle' covers all electrically propelled road vehicles with four or more wheels that are capable of accepting electrical charge from a source external to the vehicle, including pure electric vehicles, plug-in hybrid electric vehicles (PHEV) and extended-range electric vehicles (E-REV).
EV	Electric vehicle
E-REV	Extended range electric vehicle
EVSE	Electric vehicle supply equipment
GPRS	General Packet Radio Service
ICCB	In-cable control box for Mode 2 charging.
PHEV	Plug-in hybrid electric vehicle
PME	Protective multiple earthing
RCBO	Residual-current-operated circuit-breaker with integral overcurrent protection
RCCB	Residual-current-operated circuit-breaker without integral overcurrent protection
RCD	Residual current device
RFID	Radio frequency identification device
Time-of-use tariffs	Time-of-use tariffs implement a pricing strategy where the supplier of electricity may vary the price depending on the time of day that the electricity is delivered.

TMO	Traffic management order
Vehicle connector	Part of a vehicle coupler integral with, or intended to be attached to, the flexible cable connected to the AC supply network (mains).
Vehicle coupler	Means of enabling the manual connection of a flexible cable to an electric vehicle for the purpose of charging. **Note:** A vehicle coupler consists of two parts: a vehicle connector and a vehicle inlet.
Vehicle inlet	That part of the vehicle coupler that is integrated into the structure of the vehicle.
WPT	Wireless power transfer: a term for inductive charging.

ANNEX I

Figures and Tables

List of Figures

List of Tables

ANNEX J

References

The following regulations, standards and guidance are referenced within this Code of Practice:

The Electricity Safety, Quality and Continuity Regulations 2002 (as amended)

The Electromagnetic Compatibility Regulations 2016

The Electrical Equipment (Safety) Regulations 2016

The Building Regulations 2010: Part P (Electrical safety – Dwellings), 2013 Edition

BS 1363 series

BS 1363-1:2016+A1:2018 – *13A plugs, socket-outlets, adaptors and connection units. Specification for rewirable and non-rewirable 13A fused plugs*

BS 1363-2:2016+A1:2018 – *13A plugs, socket-outlets, adaptors and connection units. Specification for 13A switched and unswitched socket-outlets*

BS 4662:2006+A1:2009 – *Boxes for flush mounting of electrical accessories. Requirements, test methods and dimensions*

BS 7671:2018 – *Requirements for Electrical Installations, IET Wiring Regulations, Eighteenth Edition*

BS 8300 series

BS 8300-1:2018 – *Design of an accessible and inclusive built environment. Part 1: External environment – Code of practice*

BS 8300-2:2018 – *Design of an accessible and inclusive built environment. Part 2: Buildings – Code of practice*

BS EN 50174 series

BS EN 50174-1:2009+A2:2014 – *Information technology. Cabling installation. Installation specification and quality assurance*

BS EN 50174-2:2009+A2:2014 – *Information technology. Cabling installation. Installation planning and practices inside buildings*

BS EN 50174-3:2013+A1:2017 – *Information technology. Cabling installation. Installation planning and practices outside buildings*

BS EN 50413:2008+A1:2013 – *Basic standard on measurement and calculation procedures for human exposure to electric, magnetic and electromagnetic fields (0 Hz - 300 GHz)*

BS EN 50491-11:2015 – *General requirements for Home and Building Electronic Systems (HBES) and Building Automation and Control Systems (BACS). Smart Metering. Application Specifications. Simple External Consumer Display*

BS EN 50499:2008 – *Procedure for the assessment of the exposure of workers to electromagnetic fields*

BS EN 50500:2008+A1:2015 – *Measurement procedures of magnetic field levels generated by electronic and electrical apparatus in the railway environment with respect to human exposure*

BS EN 50664:2017 – *Generic standard to demonstrate the compliance of equipment used by workers with limits on exposure to electromagnetic fields (0 Hz - 300 GHz), when put into service or in situ*

BS EN 50665:2017 – *Generic standard for assessment of electronic and electrical equipment related to human exposure restrictions for electromagnetic fields (0 Hz - 300 GHz)*

BS EN 60309 series

    BS EN 60309-1:1999+A2:2012 – *Plugs, socket-outlets and couplers for industrial purposes. General requirements*

    BS EN 60309-2:1999+A2:2012 – *Plugs, socket-outlets and couplers for industrial purposes. Dimensional interchangeability requirements for pin and contact-tube accessories*

    BS EN 60309-4:2007+A1:2012 – *Plugs, socket-outlets and couplers for industrial purposes. Switched socket-outlets and connectors with or without interlock*

BS EN 60529:1992+A2:2013 – *Degrees of protection provided by enclosures (IP code)*

BS EN 60670-1:2005+A1:2013 – *Boxes and enclosures for electrical accessories for household and similar fixed electrical installations. General requirements*

BS EN 60947-2:2017 – *Low voltage switchgear and controlgear. Circuit-breakers*

BS EN 61000-6- and BS IEC 61000-6 series

    BS EN 61000-6-1:2007 – *Electromagnetic compatibility (EMC). Generic standards. Immunity for residential, commercial and light-industrial environments*

    BS EN 61000-6-2:2005 – *Electromagnetic compatibility (EMC). Generic standards*

    BS EN 61000-6-3:2007+A1:2011 – *Electromagnetic compatibility (EMC). Generic standards. Emission standard for residential, commercial and light-industrial environments*

    BS EN 61000-6-4:2007+A1:2011 – *Electromagnetic compatibility (EMC). Generic standards. Emission standard for industrial environments*

    BS EN 61000-6-5:2015 – *Electromagnetic compatibility (EMC). Generic standards. Immunity for equipment used in power station and substation environment*

BS IEC 61000-6-6:2003 – *Electromagnetic compatibility (EMC). Generic standards*

BS EN 61008-1:2012+A12:2017 – *Residual current operated circuit-breakers without integral overcurrent protection for household and similar uses (RCCBs) – General rules*

BS EN 61009-1:2012+A12:2016 – *Residual current operated circuit-breakers with integral overcurrent protection for household and similar uses (RCBOs). General rules*

BS EN 61558-2-4:2009 – *Safety of transformers, reactors, power supply units and similar products for supply voltages up to 1100 V. Particular requirements and tests for isolating transformers and power supply units incorporating isolating transformers*

BS EN 61851 series

BS EN 61851-1:2011 – *Electric vehicle conductive charging system. General requirements*

BS EN 61851-21:2002 – *Electric vehicle conductive charging system. Electric vehicle requirements for conductive connection to an AC/DC supply*

BS EN 61851-22:2002 – *Electric vehicle conductive charging system. AC electric vehicle charging station*

BS EN 61851-23:2014 – *Electric vehicle conductive charging system. DC electric vehicle charging station*

BS EN 61851-24:2014 – *Electric vehicle conductive charging system. Digital communication between a d.c. EV charging station and an electric vehicle for control of d.c. charging*

BS EN 61968-9:2014 – *Application integration at electric utilities. System interfaces for distribution management. Interfaces for meter reading and control*

BS EN 62056 series *Electricity metering data exchange*

BS EN 62196 series

BS EN 62196-1:2014 – *Plugs, socket-outlets, vehicle connectors and vehicle inlets. Conductive charging of electric vehicles. General requirements*

BS EN 62196-2:2017 – *Plugs, socket-outlets, vehicle couplers and vehicle inlets. Conductive charging of electric vehicles. Dimensional compatibility and interchangeability requirements for a.c. pin and contact-tube accessories*

BS EN 62196-3:2014 – *Plugs, socket-outlets and vehicle couplers. Conductive charging of electric vehicles. Dimensional compatibility and interchangeability requirements for d.c. and a.c./d.c. pin and tube-type contact vehicle couplers*

BS EN 62262:2002 – *Degrees of protection provided by enclosures for electrical equipment against external mechanical impacts (IK code)*

BS EN 62305 series – *Protection against lightning*

BS EN 62311:2008 – *Assessment of electronic and electrical equipment related to human exposure restrictions for electromagnetic fields (0 Hz - 300 GHz)*

BS EN 62423:2012 – *Type F and type B residual current operated circuit-breakers with and without integral overcurrent protection for household and similar uses*

BS EN 62752:2016 – *In-cable control and protection device for mode 2 charging of electric road vehicles (IC-CPD)*

Ethernet BS ISO/IEC/IEEE 8802-3:2017 – *Information technology. Telecommunications and information exchange between systems. Local and metropolitan area networks. Specific requirements. Standard for Ethernet*

IEC 60479-5:2007 Ed 1 – *Effects of current on human beings and livestock - Part 5: Touch voltage threshold values for physiological effects*

ISO/IEC/IEEE 8802.15-4:2010 – *Information technology. Telecommunications and information exchange between systems. Local and metropolitan area networks. Specific requirements. Wireless medium access control (MAC) and physical layer (PHY) specifications for low-rate wireless personal area networks (WPANs)*

PD CEN/CLC/ETSI TR 50572:2011 – *Functional reference architecture for communications in smart metering systems*

PD CLC/TS 50568-4:2015 – *Electricity metering data exchange. Lower layer PLC profile using SMITP B-PSK modulation*

PD CLC/TS 50568-8:2015 – *Electricity metering data exchange. The DLMS/COSEM suite. SMITP B-PSK PLC communication profile for neighbourhood networks. Including: The Original-SMITP PLC BPSK communication profile, The Original-SMITP Local data exchange profile and The Original-SMITP IP communication profile*

Energy Networks Association Engineering Recommendations:

G12/4 (2015, incorporating Amendment 1) – *Requirements for the application of protective multiple earthing to low voltage networks*

G59/3 (2015) – *Recommendations for the connection of generator plant to the distribution systems of licensed distribution network operators*

G83/2 (2012) – *Recommendations for the connection of type tested small-scale embedded generators (up to 16 A per phase)*

G100/1 (2016) – *Technical requirements for customer export limiting schemes*

INDEX

Index

D

E

H

IET Standards

Code of Practice
for Electric Vehicle Charging Equipment Installation
3rd Edition

This Code of Practice provides a clear overview of EV charging equipment, as well as setting out the considerations needed prior to installation and the necessary physical and electrical installation requirements. It also details what needs to be considered when installing electric vehicle charging equipment in various different locations – such as domestic dwellings, on-street locations, and commercial and industrial premises.

Key changes from the second edition include:

- Two completely new sections
  - Vehicles as Energy Storage
  - Integration with smart metering and control, automation and monitoring systems
- A new Annex
- A complete update to the new requirements in BS 7671:2018
- Bringing the Code in line with revised regulations and good practice

The risk assessments and checklists have also been reviewed and revised and can now also be bought separately as fillable pdf forms.

This very well established Code of Practice, supported by all the major stakeholders in the industry, is essential reading for anyone involved in the rapid expansion of EV charging points, and those involved in maintenance, extension, modification and periodic verification of electrical installations that incorporate EV charging.

www.**theiet**.org/standards

IET Standards
Michael Faraday House
Six Hills Way
Stevenage
Hertfordshire
SG1 2AY

ISBN 978-1-78561-680-8

9 781785 616808 >